Active Devices for Electronics

Active Devices for Electronics

WALTER A. SOWA

THE PENNSYLVANIA STATE UNIVERSITY

RINEHART PRESS SAN FRANCISCO

Copyright © 1971 by Rinehart Press
a Division of Holt, Rinehart and Winston, Inc.
All Rights Reserved
Library of Congress Catalog Card Number: 70–126534
SBN: 03–082796–5
Printed in the United States of America
1 2 3 4 038 9 8 7 6 5 4 3 2 1

Preface

In undertaking any textbook relating to the early phases of electronics, the author is faced with two principal choices: to cover the operating characteristics of major devices and then delve into various circuit configurations that utilize these major devices; or to deemphasize circuit applications and attempt to cover in adequate detail the characteristics of all devices in common usage. In this book the latter course was selected since it is deemed essential for the aspiring electronics technician to understand the similarities and dissimilarities of the majority of devices he might encounter, regardless of their application. Otherwise, the technician is faced with the program of constant retraining whenever a new device is encountered. Once the devices are understood, any student with a reasonably thorough background in ac and dc circuit theory should experience relatively little difficulty in understanding any circuit that combines both active and passive devices.

It is also believed that this approach permits the reader to make a more meaningful selection of a study program major——that is, industrial electronics, communications, television, and so forth.

Portions of this book are taken from the author's previous work, *Special Semiconductor Devices*, published by Holt, Rinehart and Winston, Inc., in 1968 as part of the **HRW Series in Electronics Technology**. The level of instruction is deemed appropriate for any type of training program in which emphasis is on practical understanding as opposed to extensive theoretical investigation. The only mathematical prerequisite is a working knowledge of elementary algebra. Approximations are used throughout the text to minimize confusion. This is helpful if the student is studying these devices for the first time.

Particular credit is due John M. Doyle, General Editor, Electronics Technology, HRW, for his direction and assistance in the preparation of this text which, together with four others—*Foundation for Electronics*, *General Electronics*, *Modern Industrial Electronics*, and *Modern Communications Electronics*—forms a continuum at the same practical, easy-to-understand level. At the time these five books were conceived, Mr. Doyle was employed as Manager of Research and Development for Electronic Aids, Inc., Baltimore, Md., manufacturers of education training systems in the electrical, electronics, and industrial instrumentation technologies at the senior high school, technical institute, and industrial-training levels. The texts, accordingly, reflect the specific requests of many hundreds of instructional personnel at the levels noted, combined with the extensive experience of the EAI technical staff.

Walter A. Sowa

Shavertown, Pennsylvania
January 1971

Contents

PREFACE v

1 AN INTRODUCTION TO SEMICONDUCTORS 1

1-1 Atomic Structure 1
1-2 Germanium (Ge) and Silicon (Si) Atoms 3
1-3 Conduction in a Pure Semiconductor 5
1-4 Crystals with Impurities 7
1-5 Summary 9
Questions and Problems 10

2 THE *p-n* JUNCTION 11

2-1 Formation of a *p-n* Junction 11
2-2 The Effects of Reverse-Biasing a *p-n* Junction 12
2-3 How Forward-Biasing Affects a *p-n* Junction 13
2-4 The Electrical Characteristics of a *p-n* Junction 14
2-5 Summary 16
Questions and Problems 17

3 THE JUNCTION TRANSISTOR 18

3-1 Transistor Action 18
3-2 The Relationship between Emitter Current, Collector Current, and Base Current 20
3-3 Common-Base Circuits and Alpha 21

3-4 Common-Emitter Circuits and Beta 22
3-5 Common-Collector Circuit 24
3-6 Summary 24
Questions and Problems 25

4 ELECTRICAL CHARACTERISTICS AND RATINGS 27

4-1 Junction-Diode Characteristics 27
4-2 Common-Base Characteristics 32
4-3 Common-Emitter Characteristics 34
4-4 Common-Collector Characteristics 35
4-5 Using Characteristic Curves To Determine Current Gain and Output Resistance 36
4-6 How Leakage Currents Vary with Temperature 39
4-7 Ratings of Semiconductors 41
4-8 Summary 42
Questions and Problems 42

5 BASIC TRANSISTOR AMPLIFIERS 44

5-1 The Common-Emitter Amplifier: dc Considerations 44
5-2 The Common-Emitter Amplifier: ac Considerations 51
5-3 Common-Base Amplifier 61
5-4 Common-Collector Amplifier 68
5-5 Comparing the Three Basic Configurations 76
5-6 Methods Used for dc Bias 79
5-7 DC Operating Point 86
5-8 Summary 88
Questions and Problems 88

6 THE FIELD-EFFECT TRANSISTOR 93

6-1 The Junction Field-Effect Transistor (JFET) 93
6-2 The Ampere–Volt Characteristics of a JFET 95
6-3 Measurement of Transconductance and I_{DSS} 98
6-4 The Common-Source Amplifier 100
6-5 The Common-Drain Amplifier (Source Follower) 105
6-6 Biasing the FET 107
6-7 The Insulated-Gate FET 109
6-8 Summary 111
Questions and Problems 112

7 THE UNIJUNCTION TRANSISTOR (UJT) 114

7-1 Basic Operation 114
7-2 Characteristic Curves and Specifications 116
7-3 Relaxation Oscillator 119
7-4 How a UJT is Used in a Control Circuit 123
7-5 Summary 124
Questions and Problems 125

8 THE SILICON CONTROLLED RECTIFIER (SCR) 126

8-1 Structure and Basic Operation 126
8-2 Firing Characteristics 129
8-3 Ratings and Characteristics 132
8-4 Simple Applications of SCRs 133
8-5 Series and Parallel Operation 137
8-6 Summary 137
Questions and Problems 138

9 THE DIAC AND TRIAC 139

9-1 Structure and Basic Operation 139
9-2 Trigger Modes 141
9-3 Triggering Considerations 144
9-4 AC Switching 145
9-5 Heat Sinks 149
9-6 Summary 152
Questions and Problems 152

10 LIGHT OPERATED DEVICES 154

10-1 The Nature of Light 154
10-2 Photoconductive Cells 158
10-3 Photovoltaic Cells 160
10-4 Photodiodes and Phototransistors 162
10-5 The Light-Activated SCR (LAS) 164
10-6 Summary 165
Questions and Problems 166

11 INTEGRATED CIRCUITS (ICs) 167

11-1 Advantages of ICs 167
11-2 Semiconductor ICs 169
11-3 Thin-Film ICs 173
11-4 Hybrid ICs 175

11-5 Limitations 176
11-6 Summary 177
Questions and Problems 178

12 OTHER SOLID-STATE DEVICES OF INTEREST 180

12-1 Zener Diodes 180
12-2 The *p-i-n* Diode 183
12-3 The Tunnel Diode 184
12-4 The Thermistor 185
12-5 The Varactor 188
12-6 Light-Emitting Diodes (LEDs) 192
12-7 Hall-Effect Devices 194
12-8 Summary 195
Questions and Problems 196

13 ELECTRON TUBES 198

13-1 Thermionic Emission 198
13-2 The Vacuum Diode 201
13-3 The Triode and Its Characteristic Curves 202
13-4 Tube Parameters 204
13-5 Interelectrode Capacitance and Its Effect 207
13-6 The Pentode and Its Characteristic Curves 211
13-7 Beam Power Tubes 212
13-8 Variable-Mu Tubes 213
13-9 Gas-Filled Tubes 214
13-10 The Thyratron 216
13-11 The Phototube 218
13-12 The Cathode-Ray Tube 219
13-13 Summary 222
Questions and Problems 223

GLOSSARY 225

INDEX 233

Active Devices for Electronics

1

An Introduction to Semiconductors

Initially, we shall define a *semiconductor* simply as a substance whose *conductivity* is greater than that of an *insulator* and less than that of a *conductor*. A more meaningful definition of the term semiconductor is presented later.

Today, semiconductors are the most common form of so-called *solid-state devices*. Since these devices have, for the most part, replaced vacuum tubes in all forms of electronics equipment, it is mandatory for anyone who contemplates a career in electronics to have a meaningful understanding of how and why they operate. As a first step in acquiring this needed knowledge, we shall investigate qualitatively some of the physical phenomena associated with semiconductor materials. The starting point in our journey is to look at the basic structure of all *atoms*.

1-1 Atomic Structure

As you no doubt know, all substances, whether they occur in a liquid, solid, or gaseous form, are composed of atoms. A two-dimensional drawing of an atom of *helium* is shown in Figure 1-1. This atom contains a *nucleus*, around which two *electrons* revolve in a manner similar to that in which planets of the solar system orbit around the sun. A force of *attraction*, called an *electric* (Coulomb) *force*, exists between the electrons and the nucleus and permits the electrons, which normally repel one another, to travel in orbit about the nucleus.

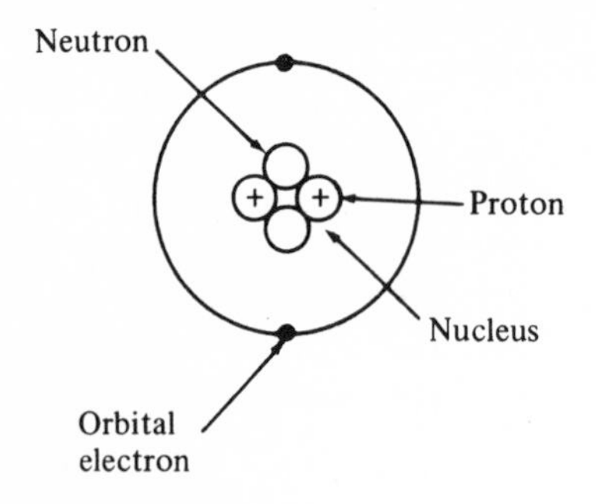

Figure 1–1 Two-dimensional illustration of an atom of helium.

Contained within the nucleus are two additional subatomic particles: *protons* and *neutrons*. The atom is in its *normal state* when the number of protons within the nucleus is *equal to* the number of orbital electrons. The number of neutrons may or may not be equal to the number of protons, but this is a matter of no consequence in our present studies.

Now, every electron and every proton possesses a distinct electrical property known as *charge*. Arbitrarily, the electron is considered to possess a *negative* charge and the proton is assumed to have an *equal positive* charge. The neutron has no charge; that is, it is electrically neutral. Since the negative charge of each electron and the positive charge of each proton are of the same magnitude (strength) but of opposite polarity, the atom, in its normal state, is electrically neutral.

It is important to realize that the electrons of every atom are *identical*. The atoms of one element are distinguished from those of another element in terms of the *number* of orbital electrons and the *number* of protons in the nucleus. If all of the elements are arranged in order of increasing mass, it is found that each element contains one more electron than its immediate predecessor on the list.

In order for an orbital electron of certain apparent mass to maintain a fixed orbit around the nucleus, it must travel at a constant velocity and develop a centrifugal force that is exactly equal to the electric force mentioned previously. Thus the electron possesses both *potential energy* (the energy of position) and *kinetic energy* (the energy of motion). To occupy a particular orbit, the electron must possess a definite *total* energy equal to the sum of the potential and kinetic energies. From this it follows that electron orbits can be discussed in terms of the *energy levels* in them. As the radius of the electron orbit increases, the kinetic energy—and, therefore, the total energy developed by the electron in orbit—also increases. From this we may conclude that the highest energy level in an atom is the orbit most remote from the nucleus and, conversely, the lowest energy level in the atom is the orbit nearest to the nucleus.

In 1925, Louis de Broglie put forth the hypotheses that all natural particles, such as electrons, possess the dual characteristics of both *particle* and *wave*, and that the orbits along which the electrons move must be an integral number of wavelengths. Now, only certain radii permit circumferences that fulfill this condition; therefore, electrons of an atom can appear only at specific energy levels.

Although no more than one electron can exist at a given energy level, it is customary to group the permissible energy levels into *shells* and *subshells*, as shown in Figure 1-2. For simplicity, only the first three shells are shown. The maximum number of electrons in a particular shell is equal to $2n^2$, where n is the shell number, counting from the nucleus.

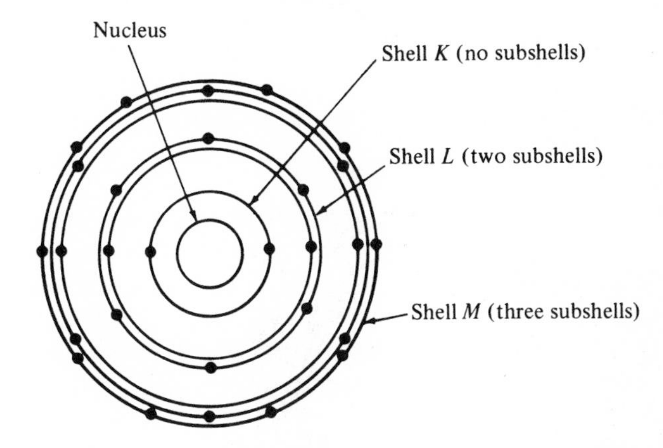

Figure 1–2 Illustrating the shell and subshell structure of atoms. The heavier atoms contain up to seven shells: *K*, *L*, *M*, *N*, *O*, *P*, and *Q*.

As the *atomic number* (that is, the number of protons in the nucleus) of the element increases, the shells begin to fill up with electrons from the lowest energy level (shell *K*) outward. This filling process continues until the first two subshells of shell *M* are filled; beyond this point, a shell may or may not be filled completely before one or two electrons appear in the next shell. In any event, however, a definite pattern is followed.

The electrons of particular interest in the study of semiconductors are those in the highest occupied energy level of the shell structure of the atom. These electrons are called *valence* electrons.

1-2 Germanium (Ge) and Silicon (Si) Atoms

Germanium and *silicon* are two materials of major interest in the fabrication of present-day semiconductor devices. Isolated atoms of these two elements are depicted in Figures 1-3(a) and (b). Notice that the silicon atom contains a total of 14 orbital electrons and the germanium atom a total of 32 orbital electrons. In each element, however, there are four valence

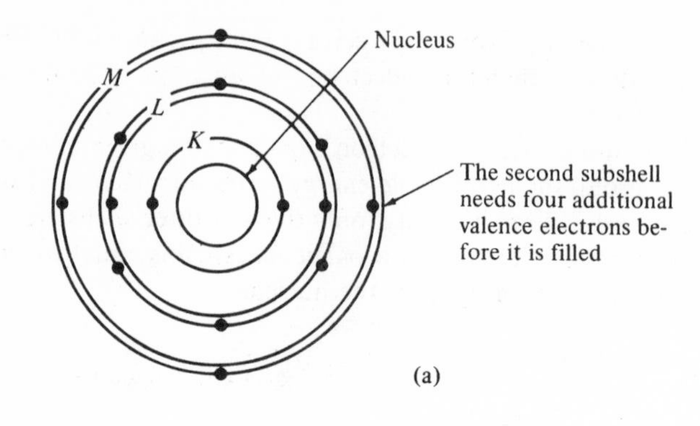

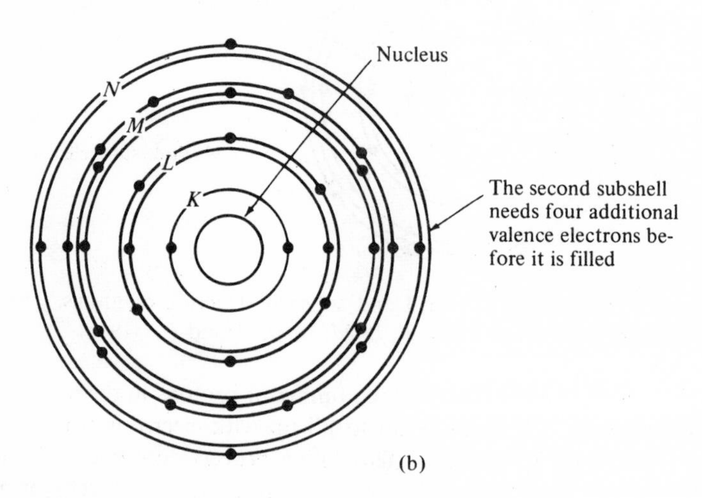

Figure 1–3 Isolated atoms of silicon and germanium.

electrons with room available for four additional electrons before the outermost subshell is filled (completed).

When Si or Ge atoms are combined with other atoms of the same type, they share each other's four valence electrons; consequently, each atom appears to have eight valence electrons that completely fill their outermost subshells. This electron sharing is called *covalent bonding* and the resulting structure is known as a *crystal*.

The concept of covalent bonding is illustrated more clearly in Figure 1-4. For improved clarity, the nucleus and all electrons, except the valence electrons of each atom, are grouped into a *core* and the valence electrons are shown without regard to their subshells. Each atom appears to have eight valence electrons in its outer shell: four of its own, plus one shared with each of its neighbors. Once each atom contains eight valence electrons in its outer

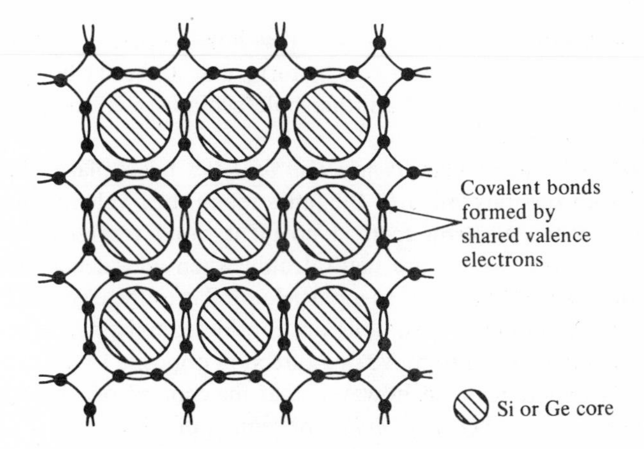

Figure 1–4 Showing the formation of a silicon or germanium crystal.

shell, the electrons are bound tightly to the atom and, *in the absence of any outside force*, cannot escape from the atom.

1-3 Conduction in a Pure Semiconductor

Assume a crystal having a structure like that shown in Figure 1-4. At a temperature of absolute zero (−273°C) every valence electron is involved in a covalent bond and is associated with its parent atom. Since there are no free electrons, the crystal is a perfect insulator.

At room temperature (about 25°C), however, the situation changes. Some of the valence electrons acquire sufficient additional energy, as a result of the application of heat, to break their atomic bonds and wander through the crystal *lattice*. (The lattice is the pattern representing the geometrical arrangement of the atoms.) Since some of the atoms are then left with a deficiency of electrons, they become *ions* and the crystal is said to be in an *ionized* state. Although we have here considered heat as the source of external energy, an electric field or incident light can also provide this ionizing energy. Thus, if an electric field is applied across the crystal, the free electrons move toward the positive terminal of the electric field source. This movement constitutes a current.

Another method of supporting a current through a pure semiconductor involves so-called *holes*. When a covalent bond between two electrons is broken, a vacancy is created in the outer shell of an atom. This vacancy is called a *hole*. Now, an electron from a nearby covalent bond may break its own bond and move to the previous vacancy. In doing so, the electron does not acquire the additional energy needed to become free in the usual sense, but rather retains its valence energy. When the electron moves from the nearby bond to the original vacancy, a new hole is created at the bond that the electron left. Another electron, in turn, may leave its bond and move into this new hole, leaving yet another hole behind. In effect, then, we have a current consisting of holes.

As noted previously, when *conduction by holes* takes place, the electrons concerned with the movement of the holes retain their valence energy. Since the absence of an electron from a covalent bond represents a localized positive charge (the atoms are neutral only when all valence electrons are present), the hole is thought of as a particle similar to the electron but having a *positive* charge. On this basis, holes are considered to be particles of definite mass (like electrons) and to move in a direction opposite to the movement of electrons. Always remember, however, that the concept of a hole is simply a concept used to describe the *jumping* movement of electrons.

To sum up, current may be conducted by two processes in a pure semiconductor. In the first process, electrons acquire enough additional energy to break their covalent bonds and be free to wander through the crystal lattice. These free electrons are then said to be in the *conduction band.* In the second process, the electrons retain their valence energy and simply move from one atom to the other; that is, they remain in the *valence band.* This constitutes a movement of holes.

When a covalent bond is broken, both an electron and a hole become available to carry current. These charge carriers are called *electron–hole pairs.* Since, in a pure semiconductor, the only charge carriers are electron–hole pairs, the number of holes is always equal to the number of electrons. The pure semiconductor is called an *intrinsic* semiconductor to distinguish it from one that contains impurities. The latter form of semiconductor, called an *extrinsic* semiconductor, is considered shortly.

Whenever an electron enters a hole, an electron–hole pair is eliminated; this process is called *recombination.* The average amount of time the charge carriers exist before recombination is called their *lifetime.*

The current conduction in a pure semiconductor is dependent upon both temperature and the electric field applied. At absolute zero there is no conduction since all of the covalent bonds are complete. As the temperature rises, small currents are produced, both from electron and hole movement, depending on the applied electric field. At excessively high temperatures a pure semiconductor may act as a conductor, since large numbers of covalent bonds are broken to supply the current carriers.

Current can occur either by *diffusion* or *drift.* Diffusion current is always in a direction to equalize a temporary, unequal distribution of charge. Drift is produced by the application of an electric field and the available charges drift in the direction of the field.

It is to be noted that, since the movement of an electron is equivalent to the movement of a hole in the opposite direction, the total current is equal to the sum of the electron and hole components.

1-4 Crystals with Impurities

The number of electron–hole pairs generated by thermal energy in an intrinsic semiconductor is generally too small to be of any practical use. To increase the number of electrons or holes available to act as charge carriers, a process called *doping* is used. In the doping process, impurity atoms are added to a pure Ge or Si crystal. The crystal is then said to be *extrinsic* rather than intrinsic.

In a typical doping process a highly refined crystal of Ge or Si is melted and a small amount of the impurity element is added. The atoms of the impurity element then diffuse through the molten Ge or Si. When the mixture cools, a new solid crystal, in which the impurity atoms are distributed among the Si and/or Ge atoms, forms.

Phosphorus (P), *arsenic* (As), and *antimony* (Sb) are three common impurity elements. Each atom of these elements has five valence electrons instead of the four associated with an atom of Ge or Si. Thus, in an extrinsic (doped) crystal to which any of these pentavalent impurities have been added, the situation illustrated in Figure 1-5 exists. Four valence electrons of the

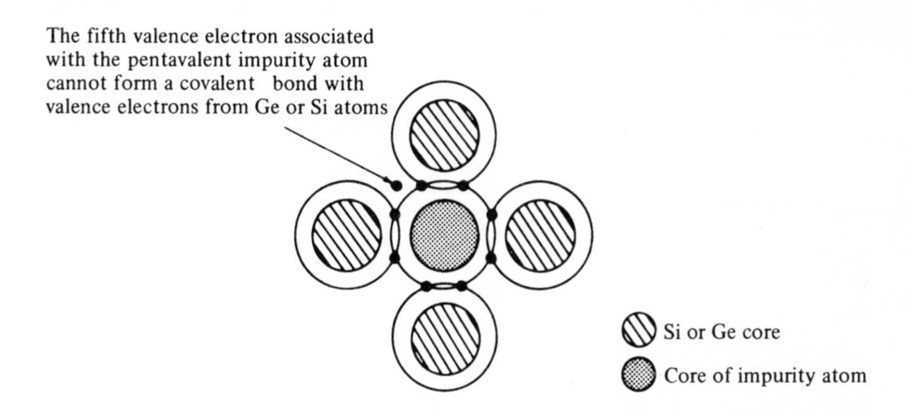

Figure 1–5 When a pentavalent impurity is added, an excess electron appears. This electron has only a weak bond with its parent atom and no bond with other valence electrons.

pentavalent atom form covalent bonds with neighboring Ge or Si valence electrons. The fifth valence electron of the pentavalent atom cannot form a bond, however, because all of the available bonds are completed; therefore, this electron is bound very loosely to its parent atom. Obviously, the number of such loosely bound electrons available in the crystal can be controlled by controlling the amount of pentavalent impurity added.

Since the addition of pentavalent atoms creates an excess of electrons in the crystal and, further, since the electrons have a negative charge, a semiconductor containing such an impurity is said to be an *n*-type semiconductor, and the *majority* carriers are electrons. Holes are *minority* carriers.

Trivalent elements—that is, elements containing three valence electrons, such as boron (B), aluminum (Al), and gallium (Ga)—are also used to dope intrinsic Ge and Si. This creates the situation depicted in Figure 1-6. The three valence electrons of the trivalent impurity atom form covalent bonds with neighboring Si or Ge atoms, but a *hole* exists in place of the electron required for a completed bond. Thus, the number of excess holes in the crystal can be controlled by controlling the amount of trivalent impurity added.

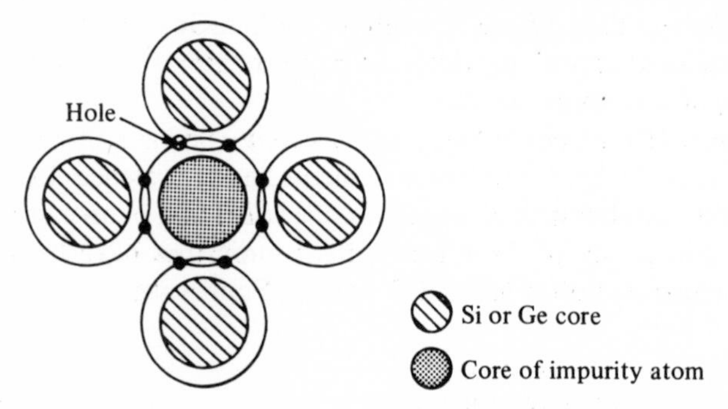

Figure 1–6 When a trivalent impurity is added, only three neighboring atoms are affected by covalent bonding. A covalent bond cannot exist with the fourth neighboring atom because of the presence of a *hole*.

When an excess of holes is created by doping, the semiconductor is said to be *p*-type material, since the hole is assumed to possess a positive charge. In *p*-type material, holes are the majority carriers and electrons the minority carriers.

Since a pentavalent impurity atom contributes an excess free electron to the crystal, it is called a *donor* impurity. An impurity that results in a *p*-type material is called an *acceptor* impurity since it accepts one electron to complete its covalent bonds.

Now, in intrinsic germanium, 0.7 electronvolt of energy (1 eV = 1.60×10^{-19} joule) is required to raise an electron from the valence band to the conduction band—that is, to create a free electron by breaking a covalent bond. In intrinsic silicon, an even greater amount of energy, 1.1 eV, is required to create a free electron. In extrinsic Ge or Si semiconductors, on the other hand, an energy of only 0.05 eV is needed to ionize donor impurities, and 0.08 eV to ionize acceptor impurities. When either type of impurity atom is ionized, of course, a charge carrier (either an electron or hole, depending on the type of impurity) becomes available for conduction. Thus, the current through a semiconductor can be controlled. This property is what makes semiconductors so useful in the field of electronics.

As you will see later, there are two major components of current: (1) that resulting from the directed movement of charge carriers introduced into the semiconductor material by doping, and (2) that resulting from the breaking of covalent bonds. The component of current attributable to the breaking of covalent bonds is undesirable and is extremely temperature-sensitive. Since a greater amount of energy is needed to break a covalent bond in silicon than in germanium (1.1 eV compared to 0.7 eV), Si is less affected by temperature increase than Ge. For this reason, silicon is used in most semiconductor applications where heat is a factor of importance. A great deal more will be said about the temperature sensitivity of semiconductors in later chapters.

1-5 Summary

Since all electrons possess a particular total energy, electron orbits are usually discussed in terms of the energy levels in them. The highest energy level in an atom is the orbit most remote from the nucleus; conversely, the lowest energy level is the orbit nearest to the nucleus.

Valence electrons exist at the highest energy level of an atom. When the atom becomes ionized—that is, either loses one or more electrons than its normal number or acquires one or more than its normal number—the ionization is caused by the movement of valence electrons.

Germanium and silicon are common semiconductor materials. Each atom of Ge or Si has room for eight valence electrons in its outermost shell; but in their normal form, only four electrons occupy this energy level. When multiple Si or Ge atoms are joined, however, there is a process of covalent bonding, where each atom shares the valence electrons of its neighbors, the outermost shell of each atom. At absolute-zero temperature, therefore, a pure crystal of Ge or Si acts like a perfect insulator. When the specimen is heated, however, some current results from the breaking of covalent bonding, but the magnitude of this current is generally too small to be of any practical use.

When impurity atoms are added to pure Ge or Si, current carriers, either in the form of free electrons or holes, become available. When the impurity is a pentavalent element, *n*-type material is produced and the material contains an excess of electrons. When the impurity is a trivalent element, *p*-type material is produced and the material contains an excess of holes.

Because it takes a greater amount of energy to break covalent bonds in silicon than in germanium, Si is used in most high-temperature devices. This reduces the undesirable component of current resulting from the breaking of covalent bonds.

Questions and Problems

1. What kind of force exists between an electron and a proton?
2. If an element contains 32 protons, how many electrons must it have to be neutral electrically?
3. What do the terms *ion*, *ionized*, and *ionization* mean?
4. Which of the electrons associated with an atom of Si are at the highest energy level? Why?
5. Into what energy band does a single semiconductor crystal group its valence electrons?
6. In what energy band do electrons exist when they are *free?*
7. In what energy band are valence electrons at a temperature of absolute zero?
8. Is it possible for conduction to occur in the valence band? Explain.
9. Does the addition of an impurity increase or decrease the number of current carriers available in a semiconductor? Explain your answer.
10. What type of semiconductor material does a so-called *donor* atom produce?
11. Why does conduction occur more readily in a doped semiconductor specimen than in a pure specimen?
12. Suppose a donor impurity is added to an intrinsic crystal of Ge. What are the majority and minority carriers in the doped semiconductor.
13. Draw a simple sketch to show how covalent bonding occurs in a pure semiconductor (Ge or Si).
14. Draw two simple sketches to show what happens to the covalent bonding in *n*-type and *p*-type Ge.

2

The *p-n* Junction

2-1 Formation of a *p-n* Junction

With great care, a single crystal can be doped to have a *p* region and an *n* region. In the formation of such a crystal, important action occurs that requires some detail study in two parts. Figure 2-1(a) shows a "stop-action" view of what the crystal looks like *prior* to any action or movement of carriers. The *p* side of the crystal has positive signs representing holes; the remaining portions of the impurity atoms are represented by circles with negative signs. The holes are in the valence band at room temperature and are the mobile particles in a *p*-type crystal. Notice that this portion of the crystal is electrically neutral.

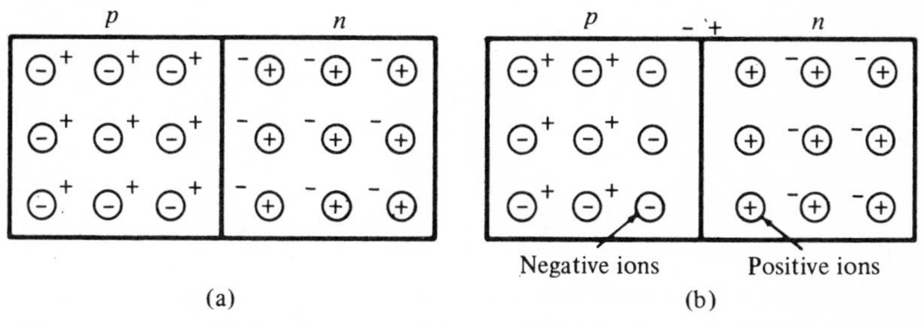

Figure 2-1 (a) Distribution of holes and electrons in a *p-n* diode *prior* to any diffusion. (b) Distribution after diffusion takes place.

On the *n* side of the crystal, the negative sign represents the fifth outer-shell electron, which is loosely attached to the remaining portion; this portion is represented by an encircled positive sign. This region is also electrically neutral.

The imaginary line that separates the two doped portions is called the junction. It must be understood that these are not two separate pieces but one crystal with two distinctly doped regions. How this is achieved is of no great consequence at this point because we are more interested in the behavior of internal charges.

The second phase of this "stop action" occurs when the junction is actually formed. In the vicinity of the junction, holes and electrons diffuse a short distance and combine with each other. Figure 2-1(b) shows the net result. The mobile holes on the *p* side combine with the mobile electrons on the *n* side, leaving a layer of immobile ions on either side of the junction. A close look at the figure indicates that the *p* region now has a layer of negative ions near the junction. Also, the *n* region has a layer of positive ions near the junction. The depth to which this combining process goes depends on the number of holes and electrons involved. In any case, more electrons from *n* region are prevented from migrating to holes in the *p* region because the electron "sees" a negative barrier near the junction (on the *p* side) by the initial combination process. It must be remembered that like charges repel, unlike attract. Conversely, mobile holes from the *p* region are prevented from reaching the electrons on the *n* side because the holes "see" a positive barrier near the junction (on the *n* side). The summation of barrier ions essentially produce a potential, referred to as a potential barrier. The region where combining of majority carriers takes place is called the depletion region, so-called because mobile carriers are absent.

Some texts call the potential barrier a potential hill. Essentially, this means that an electron in the *n* region must acquire sufficient energy (going up a hill) to overcome the barrier potential. Once it gets through the depletion region, it "slides" down the hill quite easily to a hole.

In summary, we can say that a *p-n* doped crystal has "stored" mobile carriers in each region. However, in the process of forming, a potential barrier is made at the junction that prohibits further movement of these carriers. This stalemate of events suggests that placing a voltage across the crystal is worth investigating.

2-2 The Effects of Reverse-Biasing a *p-n* Junction

Figure 2-2 shows a battery connected across a *p-n* junction. The net effect here is that the holes are drawn to the left toward the negative terminal (which attracts) and the electrons are drawn to the right toward the positive terminal. This action indeed does not promote conduction or movement of majority carriers *through* the crystal. As a matter of fact, the net effect is that the potential barrier is increased. This means that it is more difficult for electrons or holes to cross the junction. Hence, we can say that the resistivity has increased—or, conversely, that the conductivitiy has decreased—relative

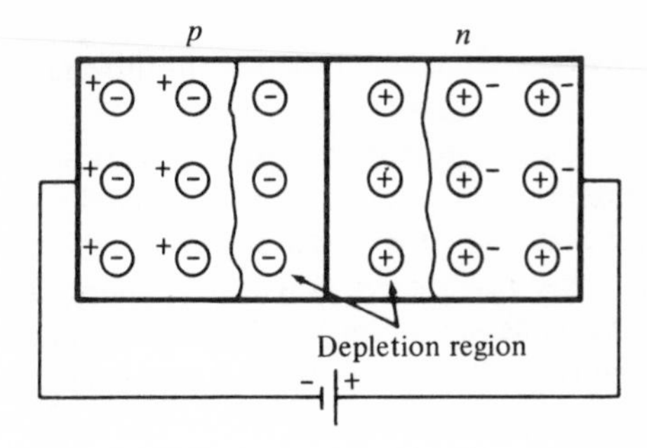

Figure 2–2 Reverse-biased *p-n* diode.

to the unbiased crystal. We can also add that moving the mobile carriers farther apart widens the depletion region. The larger we make the voltage the wider is the depletion region. Connecting a battery in this manner is referred to as *reverse-biasing*.

2-3 How Forward-Biasing Affects a *p-n* Junction

We can now anticipate the action if the battery is connected to a *p-n* junction as shown in Figure 2-3. The holes (in the *p* region) are repelled by the positive terminal of the battery toward the junction and simultaneously the electrons (in the *n* region) are repelled by the negative battery terminal toward the junction. This condition certainly enhances the likelihood of conduction because the depletion region is diminished and the potential barrier is reduced. If a low voltage is applied, holes combine with electrons *across* the junction. For each combination of electron and hole, another electron enters the *n* region from the negative terminal of the battery. Simultaneously, for each combination, a valence bond in the *p* region nearest the positive battery terminal is broken down and the liberated electron enters the battery terminal. Increasing the voltage further decreases the potential barrier and the current increases. Greater current means lower resistivity and

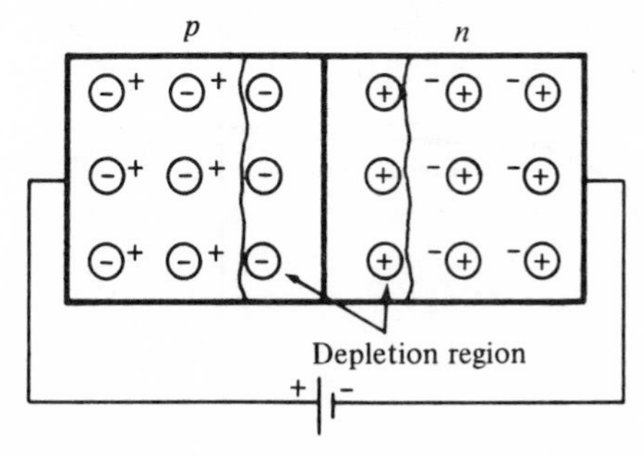

Figure 2–3 Forward-biased *p-n* diode.

increased conductivity. Connecting a battery in this manner is referred to as *forward-biasing.*

It is important to point out that if an alternating voltage were placed across a *p-n* junction, the junction would be alternately forward- and reverse-biased. This means that current would flow through the crystal and externally every half cycle. This process of converting alternating current to unidirectional current is called rectification. A two-element device that performs in this manner is called a diode. In this case, a *p-n* junction is a semiconductor diode.

2-4 The Electrical Characteristics of a *p-n* Junction

Characteristics or behavior of electric devices are usually summed up in a characteristic curve. If the device is a motor, generator, transistor, or battery, many words can be saved in explanation by portraying its characteristic curve. At the same time, the curve instantly enables us to distinguish one device from another—for example, two different transistors.

Figure 2-4 shows a generalized current–voltage characteristic for a *p-n* junction. As explained in the preceding section, in the forward-bias region, current rises rapidly as the voltage (source) is increased. Since this response has curvature, the description of this portion is said to be nonlinear. Because of nonlinearity, which is common to many electric devices, the output from such a device will not have the same shape as the input.

In the reverse-bias direction, current is quite low. Notice that the vertical scale is in microamperes for reverse-biasing and milliamperes for forward-biasing. On the basis of Section 2-3 one might challenge the fact that current,

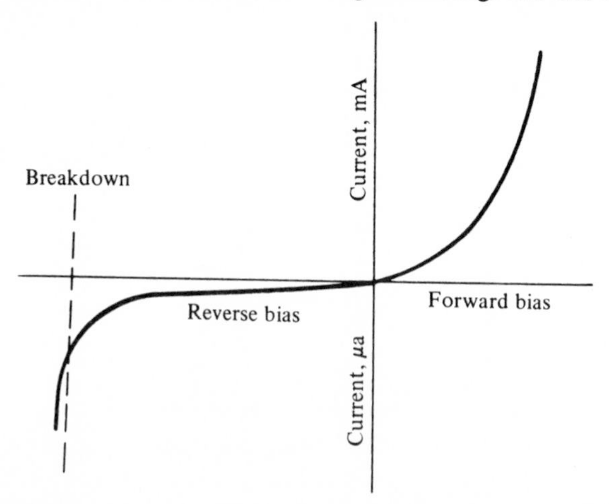

Figure 2–4 Typical semiconductor diode *I-V* characteristic.

even though minute, exists in the reverse direction. This low current flow is explained in the following manner. Because semiconductors operate at temperatures above absolute zero, electron–hole pairs are created in both regions p and n. Although the quantities are small, nonetheless they are there. Now, when a reverse-bias voltage is placed across the junction these carriers, called minority carriers, have the correct polarity to be moved *through* the junction. This means that holes in the n region are "pushed" toward the junction by the positive battery terminal and similarly electrons in the p region are "pushed" to the junction by the negative terminal of the battery. You recall that the majority carriers (contributed by the doping agents) do not contribute to this current flow. Hence, as we raise the voltage in the reverse direction, this small current increases and then remains relatively constant. But when a critical high voltage is reached as shown in Figure 2-4, breakdown occurs and excessive current flows. In rectifying applications, the useful regions of the current–voltage characteristic curve are to the right of the breakdown region. The current that flows under reverse-bias conditions is called *back current* or *leakage current*.

The foregoing discussion explains the internal behavior of a *p-n* junction. How one *p-n* junction differs from another will be explained in a subsequent chapter.

Before this chapter is concluded, it might be worthwhile to show diagrammatically how rectification is achieved. Let us refer to Figure 2-5, which is a

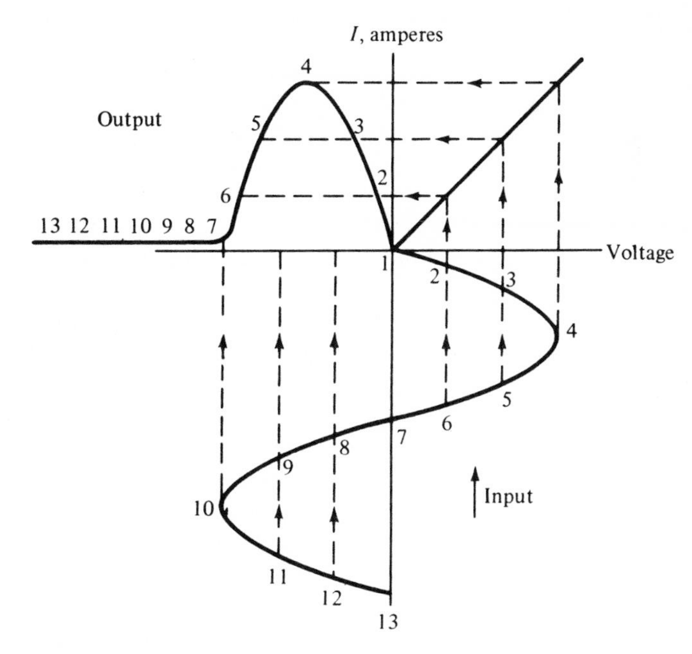

Figure 2–5 Rectification in a diode.

simplification of a junction diode characteristic. The forward characteristic is represented as a straight line (which never actually occurs) and the reverse characteristic is assumed to have zero reverse current. If a sinusoidal voltage is projected or, in the circuit, impressed across the diode, current will flow for voltages from 1 to 7. For voltages 7 to 13, no current will flow because the projection hits the characteristic at zero current flow. Putting it another way, current flows when forward bias is present and will not flow when reverse bias is present. If a resistor were placed in series with the diode the current flow through it would have the shape of the output wave. Figure 2-6 shows the schematic for the above situation. The arrowhead is the *p* side of the diode (or anode) and the solid line is the *n* side (or cathode).

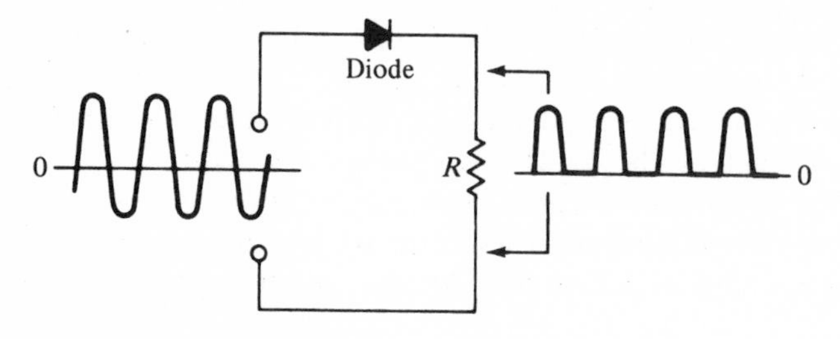

Figure 2–6 Circuit for rectification.

2-5 Summary

In Chapter 1 we learned that impurities in a crystal made the crystal a conductor when an external voltage was applied. This chapter showed how a single crystal, with *p*- and *n*-type impurities, releases holes and electrons for conduction when the voltage is in the forward direction. In the reverse direction, current flow is negligible. In one case, forward bias, majority carriers cross the junction and effectively reduce the depletion region. In the case of reverse bias, minority carriers (holes in *n* region and electrons in *p* region) cross the depletion region. Despite the movement of these minority carriers, the depletion region is larger than it is for forward bias. In a sense, the reverse-bias voltage has a greater effect on the majority carriers than on the minority carriers. Therefore, the depletion region is larger for reverse bias.

Whenever current flows relatively easily in one direction but not in another, we can say a device has a low resistance (or high conductivity) in the first direction. The converse is true in the opposite direction. Since a semiconductor diode exhibits this property, it behaves as a switch: in one direction the current flows (switch closed), and in the other direction the current does not flow (switch open). Thus, the semiconductor diode is useful as a rectifier—that is, for converting alternating current (conduction in two directions) to direct current (conduction in one direction only).

Questions and Problems

1. (True or false.) A semiconductor diode has two separate crystals joined together to form a *p-n* junction.
2. In a diode, what is the region called in which the majority carriers are removed?
3. Does forward-biasing tend to increase or decrease the depletion region?
4. (a) Does forward-biasing cause the resistivity of the diode to increase or decrease? (b) With forward-biasing, will the current increase or decrease as the voltage is raised?
5. (a) As reverse-bias voltage is increased, will the depletion region increase or decrease? (b) Reverse bias is a condition of high or low resistance?
6. What is the current flow called that flows under reverse-bias conditions?
7. Sketch the following circuit: A transformer is used to increase the ac source voltage from 110 to 550 volts; the transformer is connected to a diode and a load resistance, from which the dc voltage drop is used to feed other circuits. Indicate the polarity of the voltage drop across the resistor load.
8. If a heat lamp were placed near a diode, which current would be noticeably affected (forward or reverse)? Why?
9. Sketch a typical *I-V* diode characteristic.

3

The Junction Transistor

3-1 Transistor Action

Understanding diode action is one step away from understanding how transistors function, because the transistor is nothing more than two back-to-back diodes. In addition, with transistors we inject a new phenomenon called amplification. To appreciate transistor action, it is worth our time first to understand amplification. Any device that reasonably reproduces at its output terminals a signal which is usually larger in amplitude than the signal at its input terminals is called an amplifier.

Take, for example, the signals that appear on the antenna of an ordinary radio. A weak station, after it has been selected, must have its intelligence (voice or music) amplified before it can be audible. Vacuum tubes and transistors readily fall into this category. Or in the case of a record player, the microgrooves of the record produce small signals that, if they are to become audible, must be amplified before they reach the speakers. With this insight, the transistor action will be that much easier to perceive.

Figure 3-1 shows a single crystal with three regions doped in an *n-p-n* arrangement. These three regions are called emitter, base, and collector; in

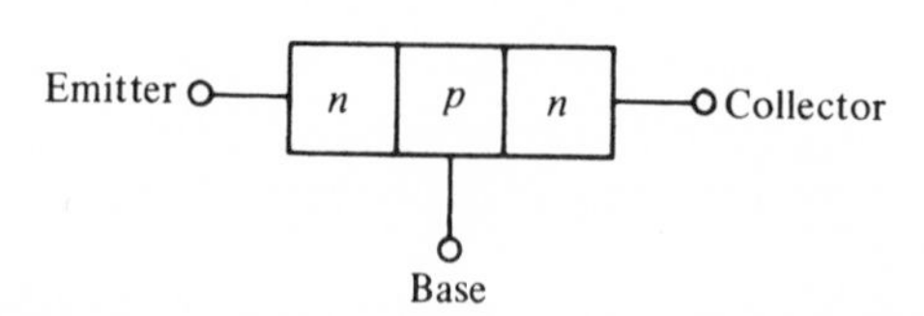

Figure 3–1 The *n-p-n* transistor.

this case, there is an *n*-type emitter, *p*-type base, and *n*-type collector. At the outset the student must understand that no transistor or tube will function unless the dc conditions are satisfied. By this, we mean that the transistor must be biased properly if it is to amplify or do anything else it is designed to do. Figure 3-2 shows that the diode formed by the emitter and base is *always* forward-biased and the diode formed by the collector and base is *always* reverse-biased.

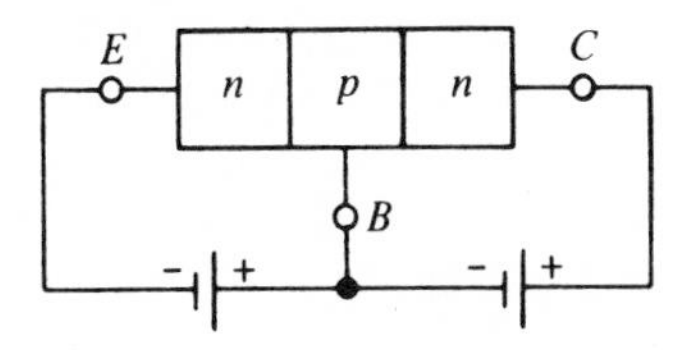

Figure 3–2 An *n-p-n* transistor with proper biases.

From the knowledge gained in Chapter 2, one can expect current flow in the forward-bias circuit and little or no current in the reverse-bias circuit. This is not so! Indeed, electrons (majority carriers in the *n* material) leave the emitter and head toward the base where there are holes and thus hopefully combine. However, the base region is usually *smaller* than the emitter region and the carrier concentration in the base (in this case, holes) is also considerably lower. The result of these two facts is that when electrons arrive in the base region their combination odds are low and these emitter-fed electrons wind up in the other *n* region, the collector. These migrant electrons now contribute to the *reverse* current, which was very small until the forward bias of the first diode released them. Perhaps this phenomenon can be understood by an analogy. Let us envision a dance hall where there are 100 boys on the east end, 20 boys on the west end, and 10 girls in the middle of the hall. At the start of the music (forward bias) the 100 boys surge to the middle of the floor in hopes of "combining." Indeed, 10 boys will combine with the 10 girls, but the momentum of the remaining 90 boys takes them to the west side where they "join" the 20 west side boys even if there is no attraction between them. Summarizing electrically, the emitter injected electrons into the base. Because the base is relatively small and has low carrier concentration, these electrons travel to the collector region. These electrons contribute to the reverse current flow. If we increase the forward bias between emitter and base, many more electrons are injected into the base and consequently the collector reverse current flow commensurately increases.

Without going into too much detail the student can easily envision some kind of weak signal varying the amount of forward bias (input) and a corresponding change occurring at the collector (output). Hopefully, amplification may occur.

It was mentioned earlier in the text that our basic aim is to find some source of electrons and then gain some measure of control over them. We have now reached that point. A more detailed study of amplification will follow later.

To this point only an *n-p-n* transistor has been discussed. We can conclude that a *p-n-p* transistor would give the same results. In this case, the emitter, which is *p*-type, injects holes into the *n*-type base where, instead of combining, most of the holes diffuse through and to the *p*-type collector. The increase in holes at the collector contributes to the initially small reverse current.

Figure 3-3 shows the transistor in schematic form. Figure 3-3(a) shows the emitter as an arrow and whenever the arrow points into the transistor, as shown, the transistor is *p-n-p*. Figure 3-3(b) shows the arrow pointing away from the transistor; this is an *n-p-n* transistor. In both diagrams the emitter-to-base diode is forward-biased and the collector-to-base diode is reverse-biased.

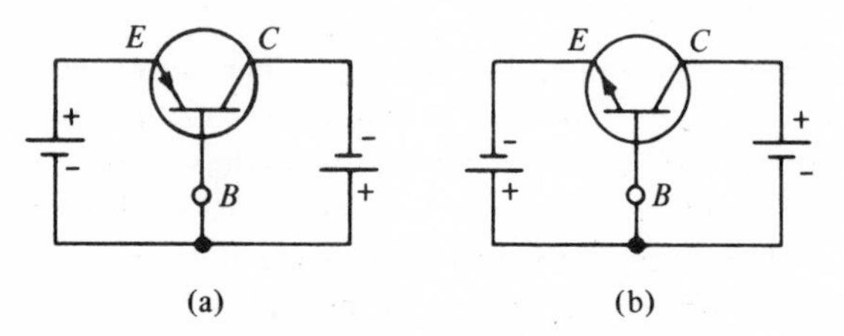

Figure 3–3 (a) A *p-n-p* transistor with proper biases. (b) An *n-p-n* transistor with proper biases.

3-2 The Relationship between Emitter Current, Collector Current, and Base Current

The three ways that transistors may be connected will be shown in this chapter. However, no matter what configuration is used, the previous paragraph implied that a relationship existed between emitter, collector, and base currents. If current-sensing meters were placed in a circuit as shown in Figure 3-4, current magnitudes would reveal that

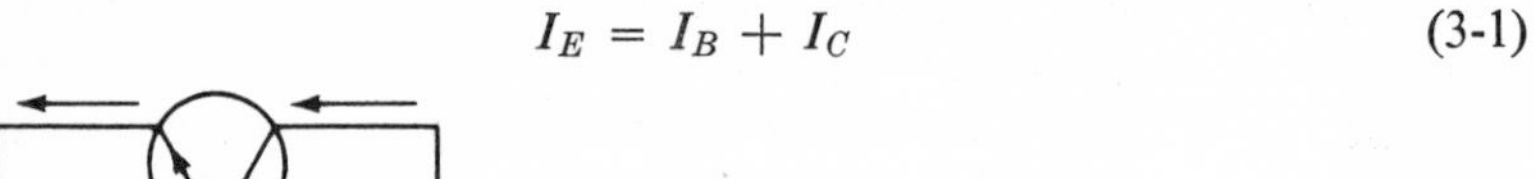

$$I_E = I_B + I_C \tag{3-1}$$

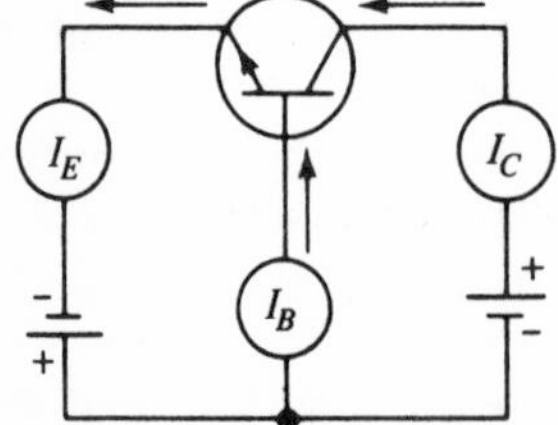

Figure 3–4 Current flow in an *n-p-n* transistor.

Since combining of holes and electrons is rather meager in the base region, I_B is rather small. Therefore, for all practical purposes,

$$I_E \simeq I_C \tag{3-2}$$

To complete the picture we can use conventional current flow (current leaving positive terminal of a battery) and indicate I_E as leaving the transistor, which conforms to the emitter-arrow direction. And since $I_B + I_C$ must equal I_E, we must show I_B and I_C as vectors entering the transistor. These vectors may seem trivial at this point but current direction will assist us later in determining voltage drops and polarities.

3-3 Common-Base Circuits and Alpha

To identify the connection configuration of a transistor it is necessary to trace where the two ac input leads go and where the two ac output leads go. One lead—base, emitter, or collector—usually winds up as being common to both input and output. The common element then determines whether we have common emitter, common base, or common collector. Figure 3-5 shows a common-base configuration. This illustration is similar to Figure 3-3 except for the addition of a generator in the emitter lead and a resistor in the collector lead. If the input is the generator its leads go to the *emitter* and *base* (assuming the battery V_{EE} has no effect on the ac voltage). If the output is considered as the ac voltage that appears across R_L, then the output leads go to the *collector* and *base* (ignoring battery V_{CC}). Since the base is common to both input and output, the circuit is a common-base configuration.

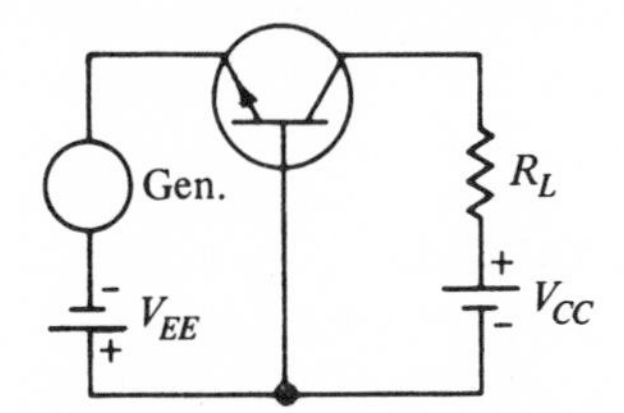

Figure 3–5 Common-base configuration.

The student may ask why and where do we use different configurations? A quick answer at this point would be that each configuration produces different characteristics and each is used where the characteristics are most suitable. This is not unlike the football coach who may use the heavy man as a lineman, a light man as halfback, and a tall man as an end. Each man has his distinctive characteristics and is used wherever the characteristics are needed.

One feature that discriminates one common base transistor connection versus another common-base transistor is the alpha (α) parameter. Essentially, it is a comparison of output current I_C versus input current I_E. When we compare dc values we obtain the dc alpha:

$$\alpha_{\mathrm{dc}} = \frac{I_C}{I_E} \tag{3-3}$$

When we compare ac values we obtain the ac alpha, which is

$$\alpha_{\mathrm{ac}} = \frac{i_c}{i_e} \tag{3-4}$$

Notice, capital letters are used for dc and lowercase letters for ac values. Often ac and dc values of alpha are almost equal. Typical values range from 0.95 to 0.99.

Students should recognize the significance of Equation (3-4). First of all, if a common-base circuit is employed, the generator varies the current in the emitter lead. It essentially changes the forward bias at the rate and amplitude of the generator. Second, the output current is the collector current and it varies at a generator rate because of the action just previously described. Hence, Equation (3-4) tells us how effective the transistor is or what percent of the carriers that leave the emitter arrive at the collector. Therefore, it is worth adding that whenever a comparison is made between ac output and ac input, we are obtaining the *gain* of a circuit or system. In this case, where we compare currents, it is current gain that is obtained.

Alpha is usually given by the manufacturer in transistor specifications. The conditions for measuring alpha require that (a) proper operating dc bias voltages are used, (b) resistor R_L in Figure 3-5 is zero, (c) currents i_e and i_c are monitored. In a practical circuit R_L is some finite value.

Comparing i_e to i_c at some value other than $R_L = 0$ produces a current gain A_i that is less than alpha. Therefore, we can consider alpha as a theoretical current gain that cannot be obtained in a practical circuit.

3-4 Common-Emitter Circuits and Beta

The student may easily verify, by using the criterion described previously, that Figure 3-6 is a common-emitter circuit. It is worth mentioning that students should also instinctively check every given circuit to verify that there is forward bias from emitter to base and reverse bias from collector to

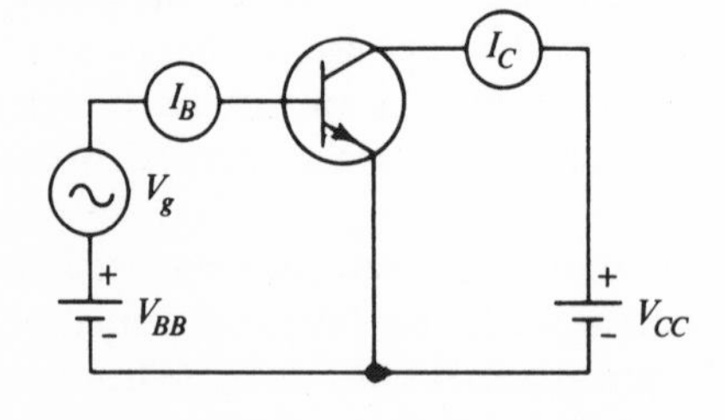

Figure 3–6 Common-emitter configuration.

base no matter what the configuration. As a review, let us analyze this circuit for appropriate dc bias. It is readily apparent that, for this *n-p-n* transistor, V_{BB}, with the indicated polarity, serves as a forward bias between emitter and base. We must assume that the ac generator V_g provides a complete path through it for the dc battery. To analyze for reverse bias between collector and base requires tracing a circuit between collector and base and summing all the voltages. There are two ways. One way is to start at the collector, go through V_{CC}, up the emitter lead and through the emitter to the base. The voltages that are encountered are V_{CC} and V_{EB}. The polarity of V_{EB} is shown on the diagram and it is opposite in polarity to V_{CC} and therefore is subtracted from it. Since V_{CC} is intentionally made greater than V_{EB} (which is usually 0.3 to 0.7 volt), the collector-to-base diode indeed is reverse-biased. The other summation technique involves starting at the collector and going through V_{CC} through V_{BB} and then to the base. Once again the polarities are opposing, and if V_{CC} is made larger than V_{BB} the collector to base is reverse-biased. This procedure may seem trivial, but later on it will be shown that V_{BB} will be replaced by voltage drops across resistors and the simple summation process just described will be of some help.

Returning to the ac considerations for Figure 3-6, we notice that the input generator is in the base lead and hence varies the base current. Also, the output current is in the collector lead. Therefore, if we compare output current versus input current, we obtain

$$\beta_{\text{ac}} = \frac{i_c}{i_b} \tag{3-5}$$

where β (beta) is a theoretical current gain. Similarly, if we compare dc values,

$$\beta_{\text{dc}} = \frac{I_C}{I_B} \tag{3-6}$$

A closer look at I_C and I_B reveals that I_C is considerably larger than I_B. As a matter of fact, I_B was so small we dropped it in proceeding from Equation (3-1), which stated

$$I_E = I_B + I_C \tag{3-7}$$

to Equation (3-2). Hence, a ratio of I_C to I_B should yield a number greater than one. Typical values that manufacturers may quote range from 10 to 800. The dc and ac β's are close to but not equal to each other. Later in this text we will show that dc values of β are used when analyzing dc conditions and ac β's are used when analyzing signal or ac conditions.

In conclusion, a relationship exists between α and β; that is,

$$\beta = \frac{\alpha}{1 - \alpha} \tag{3-8}$$

It is useful to know this relationship whenever one of these parameters needs to be converted; that is, α may be known and β required when the transistor is to be used in a common-emitter configuration.

3-5 Common-Collector Circuit

Common-collector circuits also have their distinctive characteristics. Figure 3-7 shows two ways in which a common-collector diagram may appear. In either case, the ac input appears between base and collector, and the

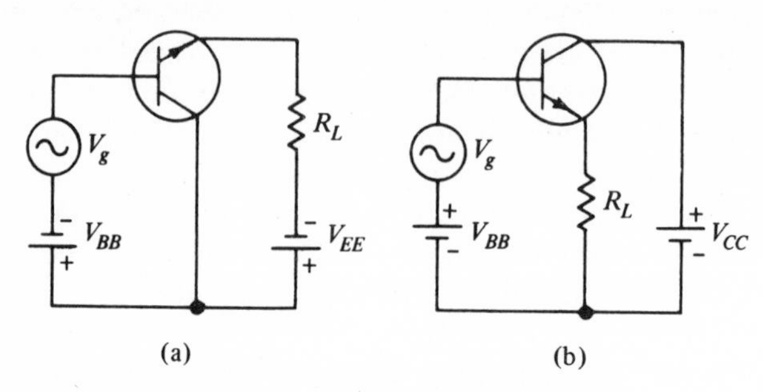

Figure 3–7 (a) and (b) Common-collector configurations.

output, which is across R_L, appears between emitter and collector. As before, we consider the batteries offering zero resistance to ac. Once again, if we compare ac output to input current, a ratio of I_E/I_B is realized. One can perceive that this ratio is slightly greater than β because β had a ratio of I_C/I_B and, from Equation (3-1), I_E is slightly greater than I_C. Therefore, for the want of an appropriate Greek letter we will call this theoretical common-collector current gain h; thus,

$$h_{\mathrm{ac}} = \frac{i_e}{i_b} \tag{3-9}$$

$$= \beta_{\mathrm{ac}} + 1 \tag{3-10}$$

$$\simeq \beta_{\mathrm{ac}} \tag{3-11}$$

Similarly,

$$h_{\mathrm{dc}} \simeq \beta_{\mathrm{dc}} \tag{3-12}$$

Returning momentarily to dc considerations, it would be appropriate for the student to determine which battery should be the larger in Figures 3-7(a) and 3-7(b).

3-6 Summary

One important aspect about transistors is that they are three-element devices and require dc potentials before any use can be realized. In every circuit the emitter-to-base diode must be forward-biased and the collector-to-base diode must be reverse-biased. In an *n-p-n* transistor, the emitter then injects electrons into the base, where hopefully they combine. Because the base is physically small and has a low hole concentration, most of the electrons wind up in the collector region, contributing current to the collector circuit. Increasing or decreasing the forward bias (simulating an ac signal) commensurately increases or decreases the collector current.

We have seen that a transistor can be connected in any of three configurations: common base, common emitter, and common collector. Each arrangement has its own characteristics. One characteristic that leads to a realization of the differences between the three is the theoretical current gain. For the common base it is α ($= i_c/i_e$), for the common emitter it is β ($= i_c/i_b$), and for the common collector it is h ($= i_e/i_b$). Alpha is usually slightly less than 1.0; beta is a number greater than 1.0; and h is approximately equal to beta.

When calculating or designing dc operating conditions, we use dc values of α, β, and h. When we make calculations relative to signal or ac voltages, we employ ac values.

Finally, the current relationship in a transistor is that emitter current equals the sum of base and collector current. For all practical purposes, the base current is relatively small and, therefore, emitter current approximately equals collector current.

Questions and Problems

1. Sketch the schematic symbol of a *p-n-p* transistor and label the elements.
2. Sketch a *p-n-p* transistor with the proper batteries attached.
3. Explain the action of a *p-n-p* transistor starting with proper biases and hole injections.
4. Determine which of the following statements are true and which are false:

   I_E is greater than I_B.
   I_E is greater than I_C.
   I_C is less than I_B.
   I_B is less than I_E.
   I_B is the smallest of all three currents.
   I_C is approximately equal to I_B.
   I_E is approximately equal to I_C.
   β is larger than α.

5. What is the mathematical ratio for α?
6. What is the mathematical ratio for β?
7. What is the equation that shows current relationship in a transistor?
8. What is the mathematical ratio for h (current gain) of a common-collector circuit?
9. Of the three theoretical current gains, which is the largest? The smallest?

10. Sketch a *p-n-p* transistor circuit in which you can expect to measure β_{dc} and β_{ac}. What values would you measure to extract these quantities?
11. Determine β if $\alpha = 0.95$.
12. Determine α if $\beta = 50$.
13. Determine I_B if $I_C = 3.50$ mA and $I_E = 3.60$ mA.
14. If these currents were measured in a common-emitter circuit, what would be the value of β_{dc}?
15. If the same currents were measured in a common-base circuit, what would be the value of α_{dc}?
16. The following data were realized in a common-emitter circuit:

Table P3-16

Test	I_B	I_C	I_E
1	100 μA	5.0 mA	5.1 mA
2	125 μA	6.0 mA	6.15 mA

Determine β_{ac}. *Hint:* $\beta_{ac} = \Delta I_C/\Delta I_B$, where Δ represents "a change in."

4

Electrical Characteristics and Ratings

4-1 Junction-Diode Characteristics

In the first three chapters we gained a basic knowledge of how a diode functions. It is now appropriate that we learn how one diode differs from another and how we can approximate the diode in a given circuit so that calculations are somewhat simplified. The student should realize that solving linear-circuit problems involving resistors and energy sources are relatively easy. The question now is: How can we find solutions if diodes are part of the circuit? The answer is to replace the diodes with circuit elements (perhaps resistors) that nearly approximate the diodes' behavior. This chapter will introduce some of these approximations.

To begin, the current–voltage characteristic of a diode in Chapter 2 is repeated, and a little deeper analysis is made. In the forward direction one can notice that no current flows until a finite value of voltage is reached and then the current rises rapidly; see Figure 4-1.

For germanium this "magic" voltage is approximately 0.3 volt; for silicon, current flows when approximately 0.7 volt is reached. These numbers are valid not only for diodes but also for transistors in which the emitter-to-base diode is forward-biased. One can readily see that with such low values of voltage required to "start" the diode, excessive currents may result and burn out the crystal if precautions are not taken. After all, the student can fully appreciate that sending high current through a small volume may cause burnout. The question is, how much is too much and what can we do to

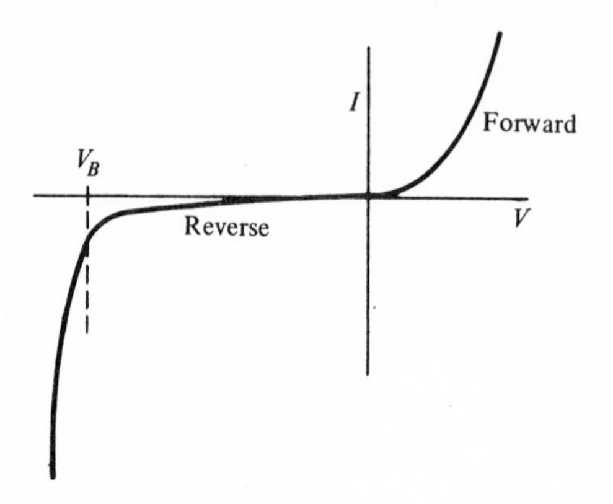

Figure 4–1 *I* versus *V* for a typical diode.

prohibit excessive current? All diodes have a maximum power rating in watts. If the manufacturer's rating is 1.0 watt, then any product combination of volts times amperes cannot exceed 1.0 watt—at least for an extended duration. As a matter of fact, specification sheets quote values of maximum allowable steady current and short-duration currents that are extremely high. However, for our purposes, let us assume that steady $V \times I$ cannot exceed the power rating given by the manufacturer.

The student must remember that all diodes require a series resistor or load of some sort to prevent excessive current flow. Figure 4-2 shows a circuit that could be used in the laboratory to obtain a forward-biased characteristic. In this case, the value of R_S is such that, when V_S is at its maximum, I_D must be less than or equal to the maximum current of the diode. Therefore, if we sum up the voltage drops around the loop,

$$-V_S + I_D R_S + V_D = 0 \tag{4-1}$$

$$R_S = \frac{V_S - V_D}{I_D} \tag{4-2}$$

EXAMPLE 1

A silicon diode is to be tested. The manufacturer's spec sheet says that the maximum current cannot exceed 500 mA. What value of R_S is required if V_S is capable of 0 to 100 volts?

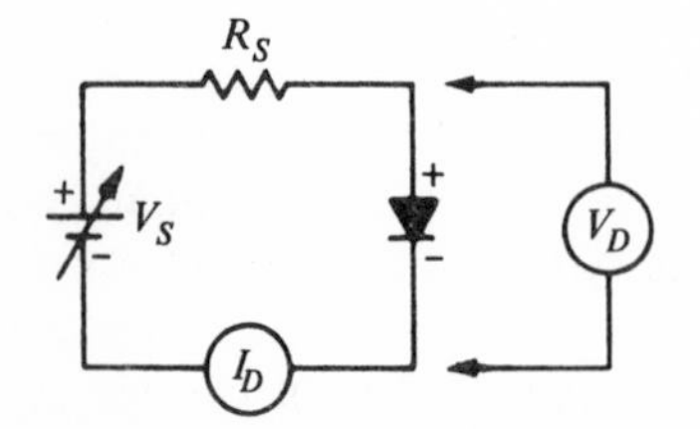

Figure 4–2 Circuit for obtaining diode *I-V* forward characteristic.

SOLUTION

$$R_S = \frac{V_S - V_D}{I_D}$$
$$= \frac{100 - 0.7}{0.5}$$
$$\simeq 200 \text{ ohms}$$

EXAMPLE 2

If the diode is energized with the preceding conditions, what is the power dissipated (a) in the diode and (b) in the resistor R_S?

SOLUTION

(a)
$$P_D = V \times I$$
$$= 0.7 \times 0.5 = 0.35 \text{ watt}$$

(b)
$$P_D = I^2 R$$
$$= (0.5)^2 \times 200 = 50 \text{ watts}$$

As a final bit of information to be extracted from the forward characteristic or diagram we can see that the arrow is the *p* material and that it points in the direction of conventional current flow. Therefore, high current values can be expected whenever the *p* material (arrow) is made positive with respect to the anode.

Analysis of the reverse-bias region of the diode (Figure 4-1) shows that very little current flows until V_B is reached. At this breakdown voltage, there is a high current flow; thus, once again a resistor must be used so that excessive current or power is not reached. However, ordinary use of diodes requires that reverse bias of this magnitude is not impressed so that this resistance consideration is not necessary. Special diodes that are designed to operate in the breakdown region are called Zener diodes. They will be covered in a later chapter.

To extract a reverse *I-V* characteristic the student may use Figure 4-2 again. However, the source must be reversed and the current meter must be replaced by a microammeter, in most cases. Also, the breakdown portion of the characteristic is not usually obtained.

We must return to the forward characteristic for a moment, because we have not yet described how a diode can be approximated by a resistance value. The student is aware that the forward characteristic is not a straight line but instead has curvature. If we take the reciprocal of the slope any place along the curve, we obtain a series of resistance values. What this means is that the diode looks like a resistor but the value of resistance changes because the slope changes. Hence, any time we wish to replace a forward-biased diode in a circuit by its equivalent circuit, we must know exactly where it is operating and then obtain the slope. However, it is not as simple as it sounds. First of all, manufacturers do not supply curve characteristics for each diode nor can we expect to run a test that will yield this curve for every diode that we may use. Thus, approximations come into play.

Figure 4-3 shows the resistance a diode offers to current flow. Figure 4-3(a) represents the resistance when the diode is well above the knee of the curve and when we lump the two resistors r_p and r_n we call it the bulk resistance, r_B. When the diode is biased near or below the knee, another resistance is added: r_j, the junction resistance. See Figure 4-3(b).

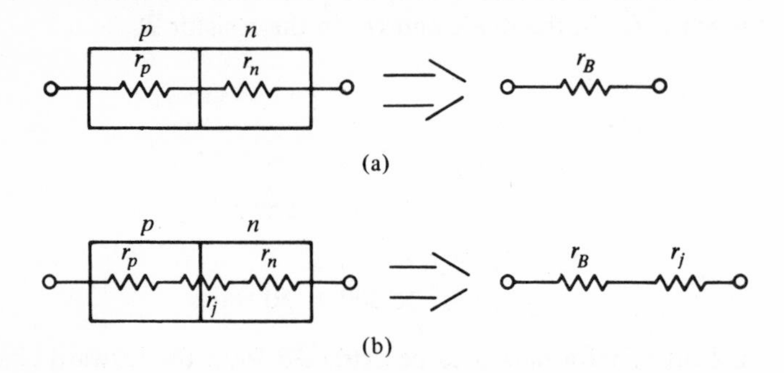

Figure 4–3 Equivalent circuit of a diode (a) well above the knee and (b) below the knee.

To approximate r_B let us examine Figure 4-4. Here a typical diode curve is approximated as straight lines. In the heavy-conduction region (above the knee), the reciprocal of the slope for a silicon diode would yield

$$r_B = \frac{\Delta V}{\Delta I} \tag{4-3}$$

$$= \frac{V_1 - 0.7}{I_1 - 0} \tag{4-4}$$

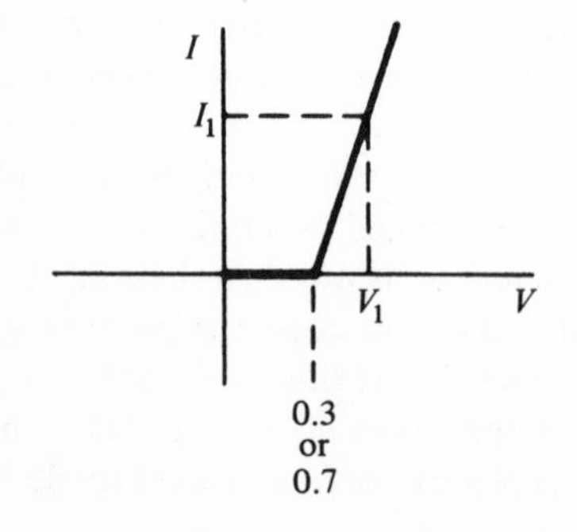

Figure 4–4 Diode approximation above the knee.

Therefore, to obtain the bulk resistance, all that is needed is one value of forward voltage and the corresponding current. This can be obtained from manufacturers' data or by one test in the laboratory.

The junction resistance r_j may be approximated as follows:

$$r_j = \frac{25 \text{ mV}}{I} \tag{4-5}$$

where I is in milliamperes.

EXAMPLE 3

A manufacturer quotes in his specifications that a germanium diode conducts 50 mA at 1 volt. Determine (a) r_B; (b) r_j for 1, 50, and 100 mA; and (c) $(r_j + r_B)$ for 1, 50, and 100 mA.

SOLUTION

(a)

$$r_B = \frac{V_1 - 0.3}{I_1}$$

$$= \frac{1\ 0 - 0.3}{(50)(10^{-3})} = 14 \text{ ohms}$$

(b)

$$r_j = \frac{25 \text{ mV}}{I}$$

$$= \frac{25 \times 10^{-3}}{1 \times 10^{-3}} = 25 \text{ ohms at 1 mA}$$

$$= \frac{25 \times 10^{-3}}{50 \times 10^{-3}} = 0.5 \text{ ohm at 50 mA}$$

$$= \frac{25 \times 10^{-3}}{100 \times 10^{-3}} = 0\ 25 \text{ ohm at 100 mA}$$

(c)

$$r = r_j + r_B$$

$$= 25 + 14 = 39 \text{ ohms at 1 mA}$$

$$= 0.5 + 14 = 14.5 \text{ ohms at 50 mA}$$

$$= 0.25 + 14 = 14.25 \text{ ohms at 100 mA}$$

Hopefully, the student will attach some significance to this example. Notice that once we got beyond the knee, r_j contributed very little to the total resistance (0.5 and 0.25 ohms—as contrasted with 14 ohms). Hence, we conclude that r_j and r_B are needed below the knee but r_B is sufficient above the knee.

A word of caution is appropriate. The use of approximations almost always precludes exact answers. This is the price we pay for expediency. The more approximate our approach the farther away will we be from the true value.

As an example, Equation (4-5) is approximate for an ideal diode at room temperature. Ideal diodes are seldom achieved and, in practice, Equation (3-4) has an upper limit of $50/I$. In other words, a diode may have a numerator that ranges from 25 to 50 in Equation (3-4). Obviously, above the knee this range does not affect the true resistance value, but below the knee it will have considerable effect. Therefore, when the student does make a theoretical calculation using approximations and then attempts to verify them in the laboratory, he can expect deviations.

A diode also has resistance in the reverse-bias direction. One can easily anticipate that this value will be relatively high, because the current is in the microampere range.

EXAMPLE 4

A manufacturer states that at 50 volts reverse bias, 10 micro-amperes flow through a diode. Determine the resistance offered by the diode.

SOLUTION

$$R = \frac{V}{I}$$

$$R = \frac{50}{(10)(10^{-6})} = 5 \text{ megohms}$$

In summarizing, it can be said that a diode can be simulated as a low resistance when it is biased in the forward direction and as a high resistance in the reverse direction. The only complication is that these resistance values depend on where on the characteristic curve the diode operates; that is, the resistances are not fixed values. This complication does not offer very much difficulty because we have the means by which we can approximate the various resistance magnitudes.

4-2 Common-Base Characteristics

As soon as a student reads or hears the words "common base" he should recall the following facts: the ac input is between the emitter and base, the output is between the collector and base, the output current is i_C, the input current is i_E, and the ratio i_C/i_E is a theoretical current gain α.

Before we look at a complete set of common-base characteristics, let us restate, somewhat briefly, the action of a transistor. Current carriers leave the emitter and instead of joining opposite charge carriers in the base, the majority of them wind up in the collector where they contribute to the *reverse-bias current flow*. We do recall that the collector-to-base voltage is reverse bias. If the emitter-to-base bias is removed then a typical reverse-bias diode characteristic would result, as shown in Figure 4-5(a). There is a very low current flow, which increases slightly as the voltage increases. If we place some forward bias on the emitter–base diode, we rely on transistor action to have more carriers at the collector. The net result is greater current flow than before as we increase V_{CB} from zero, as shown in Figure 4-5(b). If we inject still more carriers into the base by increasing the forward emitter–base bias, then I_C increases correspondingly, as shown in Figure 4-5(c). If we group these three characteristic curves we obtain what is called a family of curves.

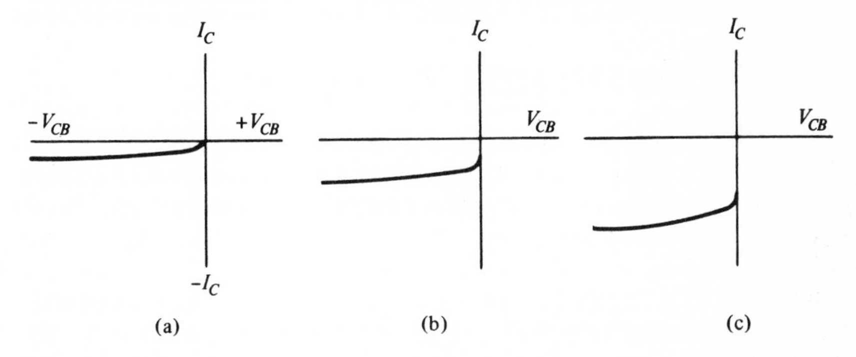

Figure 4–5 Collector current versus collector-to-base voltage when I_E equals (a) zero, (b) I_1, and (c) I_2.

If we also flip these curves into the first quadrant then we have them portrayed just as we can expect to see them in texts or transistor manuals. Figure 4-6 portrays just that, along with a circuit that can be used to extract the quantities to plot the curves. We must completely describe the conditions under which each horizontal line was obtained. In this case the lowest curve was obtained when the emitter–base voltage was zero or $I_E = 0$. The next one was achieved when I_E was some finite value I_1, and so forth with I_2 and any other curves that are obtained.

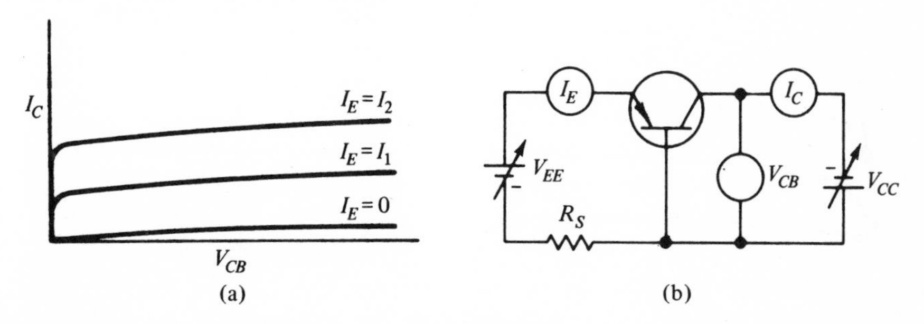

Figure 4–6 (a) Common-base output characteristics. (b) Circuit to obtain CB curves.

A close look at the coordinates of the characteristic curves will reveal that they are I_C (the output current) and V_{CB} (essentially the output voltage). Therefore, this family of curves represents the output characteristics. In the field of curves we have I_E, which is in the input circuit. Therefore, and most importantly, we have a graphical representation of how output values are affected by different input conditions. Automatically, one suspects that current gain can be extracted from these curves, because output and input currents appear. We will work on this phase in a later section.

4-3 Common-Emitter Characteristics

If the previous section is understood, then not too much difficulty will be encountered visualizing common-emitter characteristics. To recapitulate: common emitter means the input is between base and emitter, the output is between collector and emitter, input current is i_B, output current is i_C, and the ratio i_C/i_B is a theoretical current gain β.

It will not be necessary to go through the process, as we did in the previous section, of producing individual reverse characteristics and then flipping them into the first quadrant. Suffice it to say that henceforth all characteristics will be portrayed in the first quadrant. More important, what are the coordinates of the family of curves? If we conclude that an *I-V* graph must appear, then what *I* and what *V*? The common-base curves had output current and voltages of I_C and V_{CB}. Commensurately, in the common-emitter configuration output current is I_C and output voltage is V_{CE}. To obtain a family of curves we need different values of input current just as in the common-base case. This input current is I_B, as shown in Figure 4-7. For a quick verification, notice that when I_B is equal to zero, we obtain the typical reverse-bias diode characteristic. When the forward bias between emitter and base is increased,

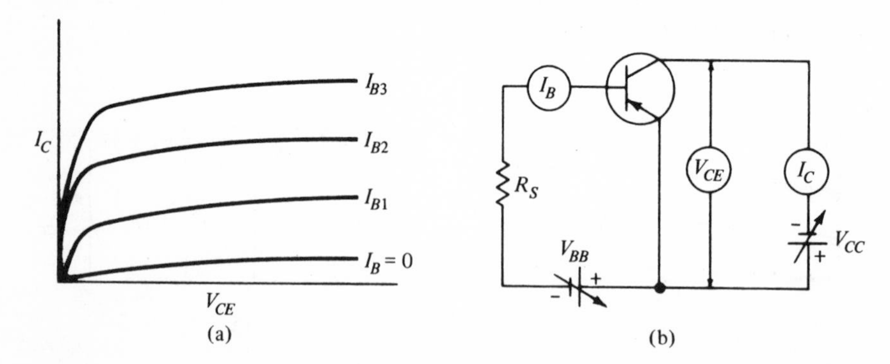

Figure 4–7 (a) Common-emitter output characteristics. (b) Circuit to obtain CE curves.

producing larger values of I_B, output currents also increase. In transistors, the range of I_C goes from milliamperes to amperes; correspondingly, I_B ranges from microamperes to milliamperes.

Before we proceed with the common-collector circuit, let us describe the mental process through which we must go so that these characteristics become relatively easy to picture. And if the results are worthwhile, we can apply them to the common collector. First of all, the characteristic must be portrayed on an I-V graph. Second, the I and V must be output values. Finally, in the field of the graph, the input characteristic must appear; this is a current parameter because the transistor is a current-controlled device.

As a review let us apply the above rules to the common emitter. Envisioning a common emitter connection tells us that the output I is I_C, the output V is V_{CE}. Notice that the output V subscripts start with the same letter as the output current and the second letter denotes the common element. Finally, curves are drawn in the field of the graph at different values of input current, which is I_B in this case.

4-4 Common-Collector Characteristics

A common-collector circuit is reproduced in Figure 4-8. We will extract common-collector characteristics by looking at this figure and at the same time use the rules we made up in Section 4-3. Hopefully, the student can anticipate and reproduce the desired characteristics without reading ahead. In any case, the output I is I_E and the output V must be V_{EC} (first letter of subscript correlates with output current). The graph must represent how I_C and V_{EC} are affected at different values of input current (in this case, I_B). Figure 4-9 is a reproduction of what we concluded.

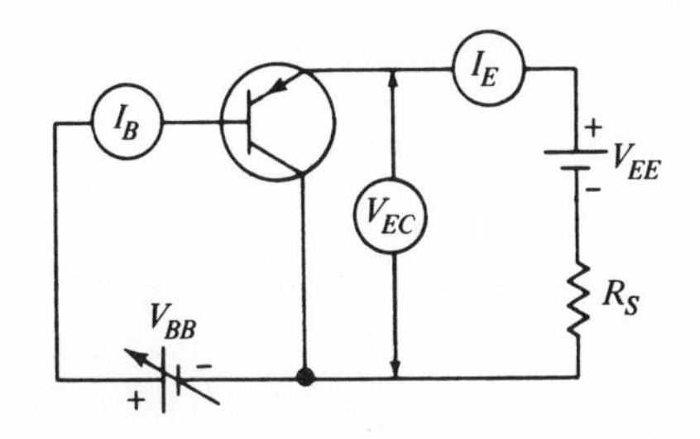

Figure 4–8 Common-collector circuit.

Notice that if we apply our approximation $I_C \simeq I_E$, knowing that $V_{EC} = V_{CE}$, the resulting family of curves becomes common-emitter characteristics. Generally, we use common-emitter characteristics whenever common-collector characteristics are required.

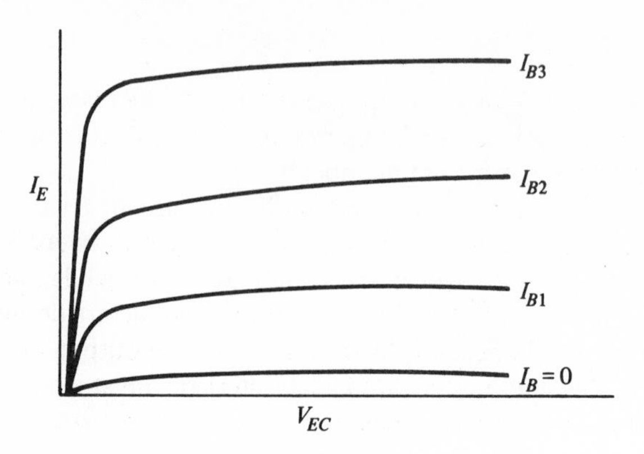

Figure 4-9 Common-collector characteristics.

4-5 Using Characteristic Curves To Determine Current Gain and Output Resistance

Earlier in the text it was pointed out that characteristic curves describe the behavior of electric devices without the use of an excessive number of words. To this we must add that we can extract from them constants that are necessary in certain mathematical equations. These equations can help us in setting up the transistor for dc operation or they can help us design and predict how the transistor will behave when it is used in a practical circuit.

We already hinted what one of these constants might be. They are α, β, and h. These can be extracted from their respective curves—that is, α from CB, β from CE, and h from CC. Chapter 3 indicated what relationships must be used to obtain them. Let us take the CE first because these characteristics lend themselves to the technique of obtaining theoretical current gain. To obtain the β_{dc} one merely picks a point on the curves and takes the ratio of I_C to I_B at that point. As an example, in Figure 4-10, I_C/I_B at point A is $5 \times 10^{-3}/30 \times 10^{-6} = 167$. To obtain β_{ac} near this point we take a ratio of changes in I_C and I_B; that is,

$$\beta_{ac} = \left. \frac{\Delta I_C}{\Delta I_B} \right|_{V_{ce}=\text{constant}} \tag{4-6}$$

The change is rather arbitrary, but it should represent small yet still readable values along a vertical line; that is, V_{CE} is held constant. If we are interested in β_{ac} near point A, we can assume a change in I_B from 20 (through point A) to 40 μA. A corresponding ΔI_C of 3.2 to 6.6 mA occurs. Therefore,

$$\begin{aligned}\beta_{ac} &= \frac{\Delta I_C}{\Delta I_B} \\ &= \frac{(6.6 - 3\ 2)(10^{-3})}{(40 - 20)(10^{-6})} \\ &= 170\end{aligned}$$

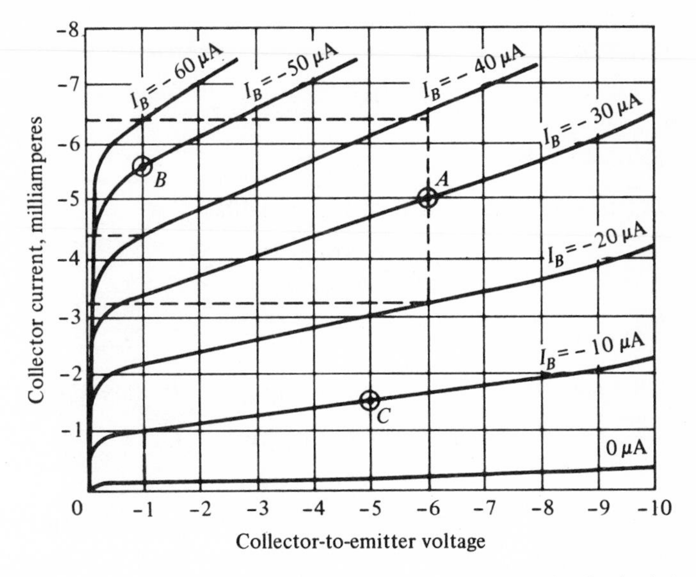

Figure 4–10 Common-emitter characteristics and method of obtaining h_{FE}, $h_{fe}$and h_{oe}

Thus β_{dc} and β_{ac} are quite close to each other. Let us investigate further by repeating the process at another point, *B*.

$$\beta_{dc} = \frac{5.6(10^{-3})}{50(10^{-6})}$$

$$= 112$$

$$\beta_{ac} = \frac{\Delta I_C}{\Delta I_B}$$

$$= \frac{(6\ 3 - 4.4)(10^{-3})}{(60 - 40)(10^{-6})}$$

$$= 95$$

Now notice that not only do β_{dc} and β_{ac} differ by a large number, but that the β values dropped from about 160 to about 100. A very important phenomenon! These ac and dc β factors are not constant but vary widely, depending upon where in the family of curves we are "operating." Manufacturers usually quote the ranges for β_{ac} and β_{dc}. A typical range of values for this transistor's β_{ac} might be 90 to 200.

We hasten to point out that β is one factor that distinguishes one transistor from another. To find more descriptive characteristics we must proceed to a factor called output conductance. It is essentially the reciprocal of the resistance that the output terminals of the transistor offer to anything that might be attached to it. To obtain the output conductance we merely find a ratio of $\Delta I_C/\Delta V_{CE}$ along a constant I_B. Hence, to obtain output conductance, we use the relation

$$G_0 = \left.\frac{\Delta I_C}{\Delta V_{CE}}\right|_{I_B=\text{constant}} \tag{4-7}$$

EXAMPLE 5

Determine G_0 at point C of Figure 4-10.

$$G_0 = \left.\frac{\Delta I_C}{\Delta V_{CE}}\right|_{I_B=C}$$

SOLUTION

If I_C goes from 1.0 to 2.0 mA along $I_B = 10$, then V_{CE} goes from 1 to 8.6 volts.

$$G_0 = \frac{(2.0 - 1.0)(10^{-3})}{(8.6 - 1)}$$

$$= 0.132 \times 10^{-3} \text{ mho}$$

$$R_0 = \frac{1}{G_0}$$

$$= \frac{1}{0.132 \times 10^{-3}} = 7.6 \text{ k}\Omega$$

It is worth mentioning that β_{ac} is much more important that G_0 because in many approximate calculations G_0 is ignored. In addition, students have heard the words hybrid parameters mentioned in relation to transistors. The two parameters we extracted from the CE curves are two of the four hybrid parameters that completely describe ac characteristics of a transistor. The remaining two cannot be obtained from the output curves because they are input characteristics. Input curves are somewhat difficult to obtain and thus we hope to circumvent the need for them. These extracted hybrid parameters will appear in a transistor manual as h_{fe} for β_{ac} and h_{oe} as output conductance. Notice the second subscript. It denotes a common-emitter configuration. The parameter h_{FE} represents β_{dc}.

We can now proceed to the other configurations with a little more ease. Since the common-collector characteristics are so much like those of the common emitter, CC values are obtained from CE curves. As stated before,

$$h_{ac} = \beta + 1 \tag{4-8}$$

or

$$h_{ac} \simeq \beta \tag{4-9}$$

Therefore, using the hybrid-parameter notation, theoretical current gain in the CC configuration is h_{fc}. Similarly, if we use CE curves to extract the output conductance for the CC configuration, $h_{oe} = h_{oc}$.

Finally, we can use CB output curves quite easily to obtain α_{dc}. We merely pick a point of interest and produce a numerical ratio of I_C/I_E. By the use of Figure 4-11, α_{dc} at point A is $1.9/2.0 = 0.95$.

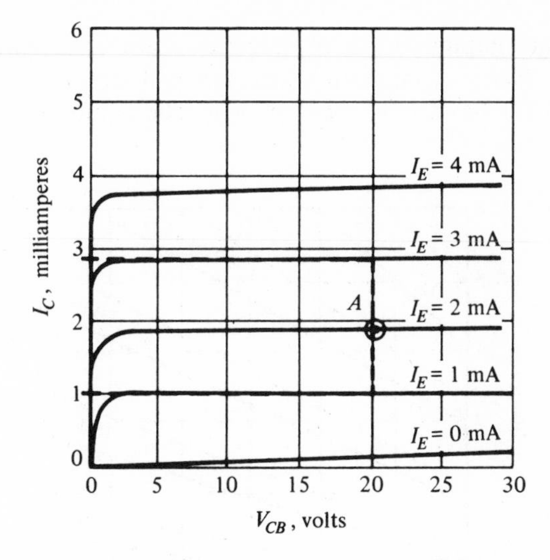

Figure 4–11 Common-base characteristics and method of obtaining h_{FB} and h_{fb}.

Obtaining α_{ac} requires V_{CB} to be constant:

$$\alpha_{ac} = \left.\frac{\Delta I_C}{\Delta I_E}\right|_{V_{CB}=\text{constant}} \tag{4-10}$$

If we choose the region about A

$$\begin{aligned}\alpha_{ac} &= \frac{(2.9 - 1.0)(10^{-3})}{(3 - 1)(10^{-3})} \\ &= \frac{1.9}{2} \\ &= 0.95\end{aligned}$$

If we repeat these two calculations in other regions, α_{dc} and α_{ac} will be quite close. Since α_{dc} is much easier to obtain we will use α_{dc} as equal to α_{ac}. Similarly, the hybrid parameter designation for α_{ac} is h_{fb} and for α_{dc} it is h_{FB}.

Obtaining output conductance is quite difficult because the slope is extremely slight. This means that h_{ob} is quite small and, conversely, R_{ob} is large. In most approximate calculations it is ignored. To prove this point the student may attempt to extract this value from Figure 4-11.

4-6 How Leakage Currents Vary with Temperature

We stated in an earlier chapter that semiconductors are temperature-sensitive. A reverse-bias diode will increase in current flow when the temperature is raised because the increased energy (heat) raises some valence-band energy electrons to the conduction-band energy level. Similarly, you will recall that output curves of transistors are essentially reverse-bias characteristics. Hence, temperature increases will increase collector-current flow. Figure 4-12 shows how each characteristic is moved upward if temperature rises.

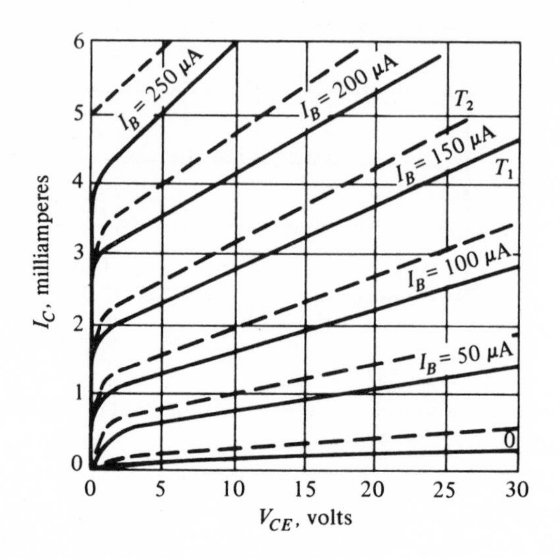

Figure 4–12 Shift in common-emitter curves when temperature increases from T_1 to T_2.

A twofold problem exists when curves are shifted this way. If T_1 characteristics are used for a design and T_2 are the actual operating conditions, then the design may not be appropriate. The proof for this is that we saw previously how β_{ac} changed as we moved from one operating point to another. In addition, if a temperature rise produces more current I_C, then increased I_C produces more heat in the transistor, and more heat produces higher current, and so on, until the transistor is burned out. This condition is called thermal runaway. We do, however, have means of combating this potential problem. Let it be sufficient to say that heat is a problem and that the leakage current is greater in a CE circuit than in a CB circuit. More specifically,

$$I_{CEO} \simeq h_{FE} I_{CBO} \tag{4-11}$$

where I_{CEO} = collector leakage current with emitter open and I_{CBO} = collector leakage current with base open. The third letter in the subscript refers to the condition of the third element of the transistor, in this case, open (O).

Summarizing, temperature changes affect transistor operation. In particular, leakage current is affected and a common-emitter circuit feels the change more than the common-base circuit by a factor of h_{FE}.

4-7 Ratings of Semiconductors

If a student opened a transistor manual, he would be somewhat confused with all the data that describe a particular transistor. We will endeavor to comprehend a few of the more important characteristics.

Transistors are rated according to power dissipation ability. Low-wattage transistors are usually small signal amplifiers, while high-wattage transistors are used as power amplifiers. No matter what purpose, the transistor wattage rating cannot be exceeded. A later text will presumably show how to calculate average power dissipated at the collector for different circuit arrangements. In any case, average power dissipated at the collector cannot exceed the given rating; that is,

$$P_C = I_C V_C \tag{4-12}$$

where P_C is the average power dissipated at the collector, and I_C and V_C are the average collector current and voltage. Wattage ratings range from a few hundred milliwatts to 50 watts.

Another rating that bears watching is the breakdown voltage, BV_{CE} or BV_{CB}. Different manufacturers use different symbols, but no matter how they are expressed these voltages represent maximum voltages that can be applied from collector to emitter and from collector to base. Usually one selects operating dc voltages well below these values.

Another maximum quantity that should not be exceeded is I_C. Depending on the transistor, I_C will range from a few milliamperes to several amperes. In design, a quick calculation can determine maximum collector current.

Manufacturers usually quote for the convenience of the designer I_{CBO} at a particular voltage. Section 4-6 gives an equation for converting to I_{CEO}. I_{CBO} is, as expected, in the low microampere range.

In addition to these important dc maxima, manufacturers sometimes will quote all four ac hybrid parameters, or perhaps only one or two. It all depends on the manufacturer. In any event these quantities may range in value by a factor of three. As an example, h_{fe} may be quoted as ranging from 40 to 130.

Perhaps the student may not realize the impact of such "loose" parameters. If a designer wished to predict the voltage gain for a given circuit, he may need h_{fe} in his equation. If the h_{fe} of the transistor he chose ranged from 40 to 130, what value should he use?

The only exact way out of this dilemma is for the designer to measure h_{fe} himself at the intended operating point.

We would be remiss if diode ratings were not mentioned. Similar to transistors, there is a maximum power and reverse voltage (PRV) that a given diode can withstand. This is, as the words imply, a reverse-biased voltage. This rating keeps the diode operating above the breakdown region. Also, in the reverse direction, manufacturers quote the amount of leakage current that flows at a particular reverse voltage. This quantity is of no great significance other than to give us a feeling for the magnitude of diode resistance in the reverse direction. Finally, maximum forward current is given. This current is the *average* forward current. This can range from milliamperes to a significant number of amperes.

4-8 Summary

In this chapter we gained some insight on the practical discriminating characteristics of semiconductor diodes and transistors. We observed that a diode, by virtue of its *I-V* relationship, has a resistance that is relatively low in the forward-bias direction and high in the reverse-bias direction. In the forward-bias region the resistance consists of two parts: block resistance and junction resistance. The sum of these two equal the total resistance. Below the knee, r_j is quite significant, but above the knee r_j is relatively small.

The transistor has three basic configurations—CB, CE, and CC—with each offering its own distinctive characteristics. One feature is the theoretical current gain. For the CB it is α, for the CE it is β, and for the CC the theoretical current gain h is approximately equal to β. In hybrid-parameter notation, the current gains are h_{fb}, h_{fe}, and h_{fc}, respectively. The term "common" as used in these configurations refers to the element that is common to both ac input and ac output.

We also observed that graphical representation of transistor behavior gives us a feeling of how one transistor differs from another. This family of curves depicts output quantities for CB, CE, and CC configurations. The coordinates of the graphs show how these output values change with changes in the input values that appear in the field of the graph.

From these curves we are able to obtain α_{dc}, α_{ac}, β_{dc}, and β_{ac}. By definition, α is a ratio of I_C to I_E, and β is a ratio of I_C to I_B. If a given transistor is connected in a CB configuration, the practical current gain will be less than 1, whereas if the same transistor is used in a CE circuit, the net practical current gain will be greater than 1.

Questions and Problems

1. What forward voltage is required across a silicon diode to make it conduct? Across a germanium diode?
2. The following data are given for a silicon diode: maximum forward current = 50 mA, reverse current = 5μA at 50 volts, and forward current = 10 mA at 1.0 volt. At 10 mA determine (a) r_j, (b) r_B, (c) $r_j + r_B$, (d) the value of resistor in series with a diode to protect the diode when placed across 50 volts, and (e) the resistance the reverse-bias diode offers at 5 μA.
3. A germanium diode is to be tested in the laboratory. The instructor limits the forward current to 100 mA and the reverse voltage to 200 volts. (a) What value of resistance is needed in series if the source voltage will

reach 200 volts? (b) What power rating should this resistor have? (c) What power can we expect to dissipate in the diode under full current conditions?

4. When the same diode was tested in the laboratory, 1.1 volts was measured when 100 mA of forward current flowed and 25 μA when 200 reverse volts were impressed. Determine: (a) the total resistance offered by the diode at 100 mA and (b) the resistance offered by the diode at 25 μA.

5. If the same germanium diode is operated at 5 mA, what is the total resistance offered by the diode? Remember, at 1.1 volts there was a current of 100 mA flowing in forward bias.

6. Sketch the CB, CE, and CC configurations.

7. What currents—I_E, I_C, or I_B—represent the input currents in CE, CB, and CC?

8. What currents represent the output currents in CE, CB, and CC configurations?

9. What current ratios determine (a) alpha, and (b) beta?

10. What is the formula that relates alpha and beta?

11. Sketch a figure that can be used in a laboratory to obtain CE characteristics. Label all meters.

12. Sketch (a) a CB family of curves and (b) a CE family of curves.

13. Determine β_{dc} and β_{ac} from Figure 4-12 at the operating point (a) $V_{CE} = 12.5$ volts and $I_C = 3$ mA, and (b) $V_{CE} = 5$ volts and $I_C = 3.6$ mA.

14. (a) Determine α_{dc} and α_{ac} from Figure 4-11 at the operating point, $V_{CB} = 5$ volts and $I_E = 3.0$ maA. (b) Convert these α's to β's.

15. What current ratio determines (a) h_{fe} and (b) h_{fb}?

16. Using Figure 4-13, determine h_{fe} at the following points: $I_B = 0.8$ mA and $V_{CE} = 5, 15, 25, 35$ volts.

17. Convert the first beta in Problem 16 to alpha.

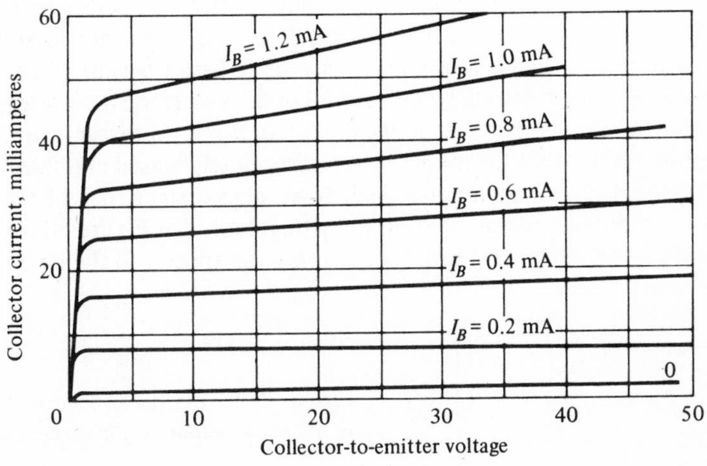

Figure 4–13

5

Basic Transistor Amplifiers

5-1 The Common-Emitter Amplifier: dc Considerations

The common-emitter amplifier is the most frequently used transistor amplifier. To comprehend how it functions and how to analyze it, it will be necessary to approximate an equivalent circuit for it and then use ordinary circuit theory. The student should not approach this section with any qualms if he has any confidence in circuit theory. The approximations will be in two parts: ac and dc equivalent circuits.

As mentioned earlier, before an amplifier can function one must take care of the dc conditions. By this is meant that emitter–base must be forward-biased and collector–base reverse-biased. It is also important to remember that forward-bias diodes require a resistance in series because high current can result once we exceed 0.3 volt (for Ge) or 0.7 volt (for Si) of forward bias. With this in mind, Figure 5-1 is shown for an *n-p-n* transistor in a CE configuration. Notice that the transistor is appropriately biased and that conventional current directions are assumed. If we placed signs for voltage drops, the loop involving the base and emitter would look like Figure 5-2(a), where 0.7 and 0.3 volt (Si or Ge) is dropped across the emitter to the base and the

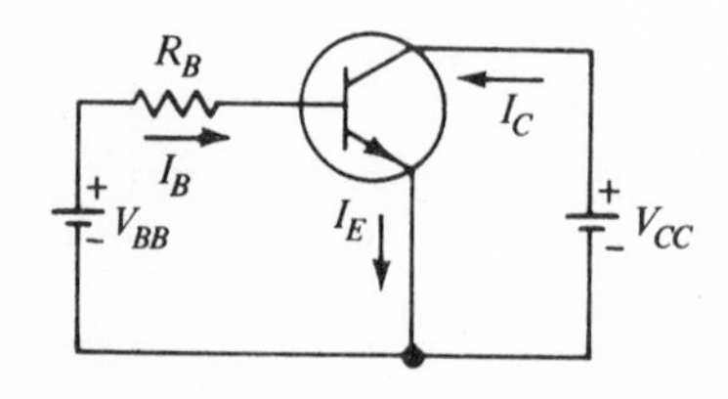

Figure 5-1 Biasing and current flow in a common-emitter configuration.

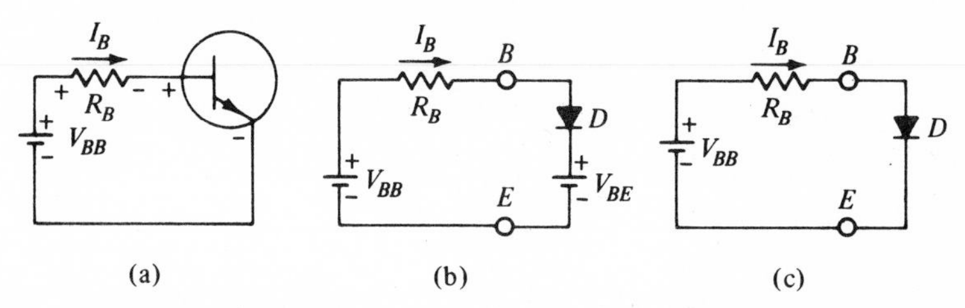

Figure 5–2 (a) Forward-biased base–emitter junction. (b) Equivalent circuit. (c) Simplified equivalent circuit.

remainder of V_{BB} across R_B. This is merely a simple series circuit and the student can readily conclude that

$$-V_{BB} + I_B R_B + V_{BE} = 0 \tag{5-1}$$

or

$$R_B = \frac{V_{BB} - V_{BE}}{I_B} \tag{5-2}$$

The student should also recognize this equation as being similar to the one used in the chapter on diodes and used to determine value of resistance required to limit current in a diode. Figure 5-2 shows how the base-to-emitter circuit is converted to an equivalent circuit. Diode D is an ideal diode, meaning zero ohms when forward biased, and V_{BE} is the voltage required for V_{BB} to overcome (0.7 or 0.3) before current flows. We can "equivalentize" a step further and throw V_{BE} away because it may not be significant. This then simplifies the transistor so that it looks like Figure 5-2(c). Thus, Equation (5-2) becomes

$$R_B \simeq \frac{V_{BB}}{I_B} \tag{5-3}$$

Perhaps this is a good place to demonstrate the cautions that must be taken when we attempt to approximate. Remember that the more we approximate, for ease of solution, the more accuracy we sacrifice.

EXAMPLE 1

A silicon transistor is used in Figure 5-1. If V_{BB} is 4 volts, and $I_B = 50\ \mu\text{A}$, determine, R_B with (a) Equation (5-2) and (b) Equation (5-3).

SOLUTION

(a)

$$\begin{aligned} R_B &= \frac{V_{BB} - V_{BE}}{I_B} \\ &= \frac{4.0 - 0.7}{50(10^{-6})} \\ &= 66\ \text{k}\Omega \end{aligned}$$

(b)

$$R_B \simeq \frac{V_{BB}}{I_B}$$

$$= \frac{4.0}{50(10^{-6})}$$

$$= 80 \text{ k}\Omega$$

From the above, we can conclude that the second solution might be too approximate. Actually, it contains a 21 percent error. Now let us take another example.

EXAMPLE 2

Assume that the same transistor is used, but in this case $V_{BB} = 12$ volts.

SOLUTION

(a)

$$R_B = \frac{V_{BB} - V_{BE}}{I_B}$$

$$= \frac{12 - 0.7}{50(10^{-6})}$$

$$= 226 \text{ k}\Omega$$

(b)

$$R_B = \frac{V_{BB}}{I_B}$$

$$= \frac{12}{50(10^{-6})}$$

$$= 240 \text{ k}\Omega$$

Here we have only a 6.2 percent discrepancy and thus solution by part (b) is acceptable. Hence the lesson we learned here is that when the battery voltage is much greater than the base–emitter voltage drop then the approximation as shown in Figure 5-2(c) is appropriate. In our case, a 12-volt value for V_{BB} can be considered *much greater than* V_{BE} (which is 0.7 volt). The student might ask, how can this "much greater than" condition be identified? A general rule to follow in all electric circuits is that whenever a factor of 10 or more is involved (12/0.7 is greater than 10), approximations can be made. The student is encouraged to prove this rule by applying it to two resistors in parallel, where one is 1000 ohms and the other is 100 ohms. Can you approximate the answer before applying the formula for two resistors in parallel?

Now let us return to the dc conditions at the output side of the transistor of Figure 5-1. Figure 5-3(a) shows this circuit; since there are no resistors in series, V_{CE} equals V_{CC}. As described previously, characteristic curves as shown in Figure 5-3(b) are produced. If we try to approximate these curves,

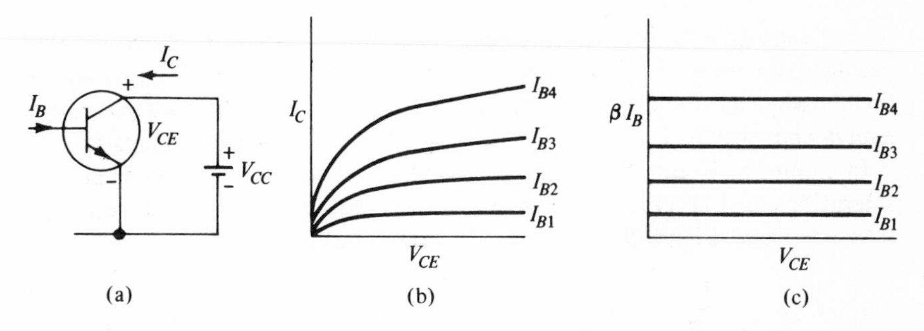

Figure 5–3 (a) Reverse-biased collector–emitter junction. (b) Typical output characteristics. (c) Approximated output characteristics.

Figure 5-3(c) results. Notice that the slopes are removed and that I_C is replaced by $\beta_{dc}I_B$. The reason for this is that if

$$\beta_{dc} = \frac{I_C}{I_B} \tag{5-4}$$

then

$$I_C = \beta_{dc}I_B \tag{5-5}$$

A close look at Figure 5-3(c) tells us that no matter what V_{CE} we may have on the transistor (say with I_{B1}), the same collector current results. Repeating this in another way, a constant collector current results for any value of collector voltage. This is a description of a constant-current generator and indeed a worthwhile approximation for the collector side of the CE amplifier.

To complete the amplifier collector circuit we must place a resistor in series with the collector. We use this resistor for ac considerations but, since it affects the dc conditions, it must be investigated. Figure 5-4(a) shows a load resistance in the collector circuit and the resulting dc equation is

$$-V_{CC} + I_C R_L + V_{CE} = 0 \tag{5-6}$$

or

$$V_{CC} = I_C R_L + V_{CE} \tag{5-7}$$

A close look at Equation (5-7) tells us that the sum of the drops across R_L and the collector-to-emitter function equals the battery supply V_{CC}. Putting

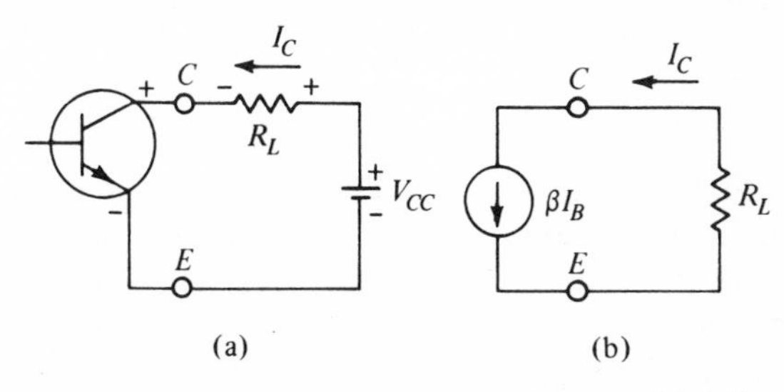

Figure 5–4 (a) CE configuration with load resistance. (b) DC equivalent circuit.

this in another way, if $I_C R_L$ increases, V_{CE} must decrease an equal amount because Equation (5-7) must be maintained. This point will be discussed further when ac conditions are studied. Figure 5-4(b) shows the output circuit equivalence which is a constant-current generator feeding a load resistor.

In summary, Figure 5-5(a) shows a typical CE amplifier with only dc considerations met, Figure 5-5(b) shows the dc equivalent circuit of an *n-p-n* transistor, and Figure 5-5(c) shows the complete circuit.

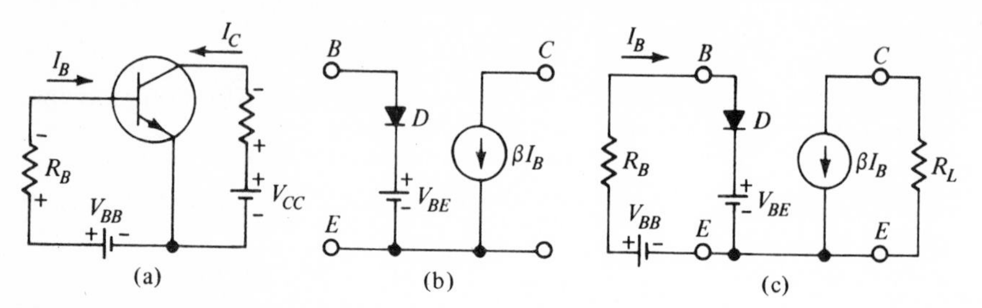

Figure 5–5 (a) CE amplifier. (b) Transistor equivalent circuit. (c) CE amplifier dc equivalent circuit.

Notice that the equivalent circuit can be easily analyzed; hence the transistor's behavior can be understood and studied. Notice also that although a complete circuit does not exist between the left side of the transistor and the right side, there does exist a relationship between their currents; that is, output direct current equals β times input direct current.

One short explanation may be appropriate concerning what kind of direct currents we are talking about and how we choose them. For small signal amplifiers, I_B is usually in the microampere range and I_C in the milliampere range. For power amplifiers, I_B is usually in the milliampere range and I_C in the ampere range. The values that are chosen obviously depend on the transistor used and on where in the field of the CE characteristics one chooses to operate. For the present let us accept the assumption that we usually choose approximately the middle of CE characteristic curves, from which we obtain the I_B, V_{CE}, and I_C required.

To solidify the knowledge gained in setting a transistor into operation, proceed through the following examples:

EXAMPLE 3

Assume that a circuit as shown in Figure 5-5(a) is to be used. An *n-p-n* silicon transistor is chosen with $V_{BB} = 6.0$ volts and $V_{CC} = 12.0$ volts. If the transistor has a B_{dc} of 50 and an I_B of 50 μA for an operating condition, determine (a) R_B, (b) I_C, (c) the dc voltage drop across R_L if R_L equals 1.0 kΩ, and (d) V_{CE}.

SOLUTION

Using Equation (5-2), we get

(a)
$$R_B = \frac{V_{BB} - V_{BE}}{I_B} = \frac{6.0 - 0.7}{50(10^{-6})} = 106 \text{ k}\Omega$$

Using Equation (5-4), we get

(b)
$$I_C = \beta I_B = 50(50)(10^{-6}) = 2.5 \text{ mA}$$

(c)
$$V_{R_L} = I_C R_L = 2.5(10^{-3})(1.0)(10^3) = 2.5 \text{ volts}$$

By the use of Equation (5-6),

(d)
$$V_{CE} = V_{CC} - I_C R_L = 12 - 2.5 = 9.5 \text{ volts}$$

EXAMPLE 4

The circuit shown in Figure 5-5(a) was tested and the following dc quantities were measured with a voltmeter: $V_{CE} = 5.0$ volts, $V_{CC} = 10$ volts, $V_{BB} = 1.5$ volts. If R_L is 5 kΩ, R_B is 100 kΩ, and we assume that the transistor is germanium, determine (a) I_C, (b) I_B, (c) B_{dc}.

SOLUTION

(a)
$$V_{R_L} = V_{CC} - V_{CE} = 10 - 5 = 5.0 \text{ volts}$$

$$I_C = \frac{V_{R_L}}{R_L} = \frac{5.0}{5(10^3)} = 1.0 \text{ mA}$$

(b)
$$I_B = \frac{V_{BB} - V_{BE}}{R_B}$$
$$= \frac{1.5 - 0.3}{100(10^3)}$$
$$= 12\ \mu\text{A}$$

(c)
$$\beta_{\text{dc}} = \frac{I_C}{I_B}$$
$$= \frac{1.0(10^{-3})}{12(10^{-6})}$$
$$= 83$$

EXAMPLE 5

An *n-p-n* silicon transistor is used in a circuit as shown in Figure 5-5(a). By employing a set of characteristics an operating point was chosen such that $I_B = 100\ \mu\text{A}$, $I_C = 4.0$ mA, and $V_{CE} = 5.0$ volts. Determine; (a) β_{dc}, (b) the value of R_L if V_{CC} is 12.0 volts, and (c) the current through R_L if R_L is decreased.

SOLUTION

(a)
$$\beta_{\text{dc}} = \frac{I_C}{I_B}$$
$$= \frac{4.0(10^{-3})}{100(10^{-6})}$$
$$= 40$$

(b)
$$V_{R_L} = V_{CC} - V_{CE}$$
$$= 12 - 5.0$$
$$= 7.0 \text{ volts}$$
$$R_L = \frac{V}{I}$$
$$= \frac{7.0}{4.0(10^{-3})}$$
$$= 1750 \text{ ohms}$$

(c) Since the transistor is a constant current source,
$$I_C = 4.0 \text{ mA}$$

5-2 The Common-Emitter Amplifier: ac Considerations

Once a transistor is set into operation by satisfying the dc requirements, it becomes necessary to find a way to inject the ac signal and a means to send it on to the next stage. Figure 5-6(a) employs a transformer coupled input not only to demonstrate how ac is injected but to assist us in explaining that voltage gain is indeed achieved. First note that we have the same circuit that was used in satisfying the dc conditions except that we interrupted the

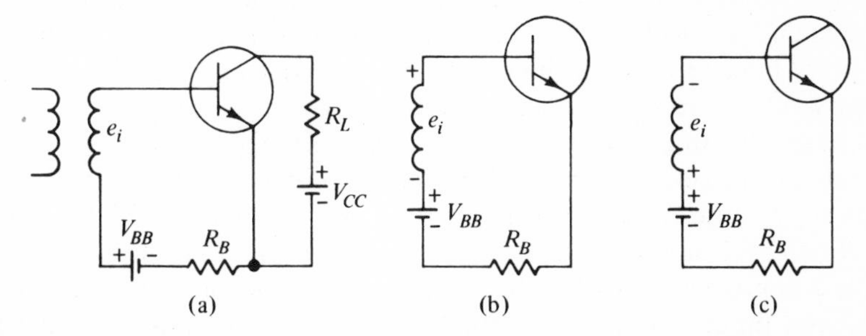

Figure 5–6 (a) CE amplifier with signal input. (b) Input circuit with signal-voltage-aiding dc bias. (c) Input circuit with signal-bucking dc bias.

input circuit with a secondary winding of a transformer. This means that whatever voltage appears across this winding is added to or subtracted from (depending on the instantaneous ac voltage) the dc voltage. We pause here to explain because this may be the first time the student has encountered a circuit in which ac and dc appear simultaneously. We can envision what happens by imagining two batteries in series, each 1.5 volts except that one is adjustable to cover the range from zero to 1.5 volts. This is shown in Figure 5-7(a). The net voltage output can have any value from 1.5 to 3.0 depending on what value the adjustable source is set at. In Figure 5-7(b), where the batteries buck each other, the voltage from *A* to *B* may range from zero to 1.5 volts. In either case, if the variable voltage is "wiggled," the output voltage, *A* to *B*, is simultaneously wiggled. This is similar to what appears in Figure 5-6 except that the voltage across the transformer wiggles itself at a

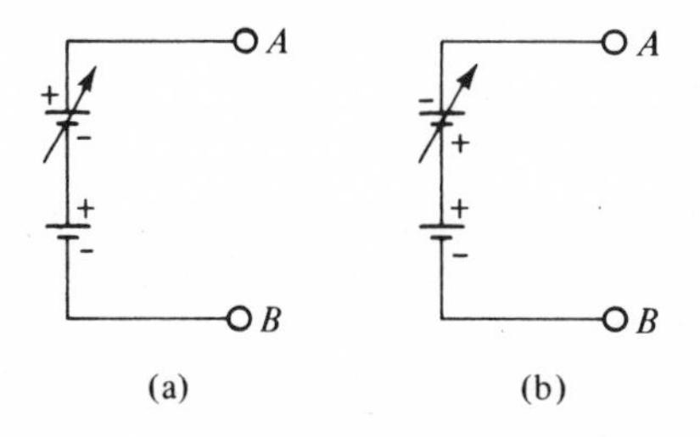

Figure 5–7 (a) Batteries aiding. (b) Batteries bucking.

signal rate. If we assume that the transformer is momentarily positive at its upper end, as in Figure 5-6(b), then the two sources are aiding each other and producing a larger forward bias. This causes more carrier to leave the emitter and eventually reach the collector. On the other hand, if the signal reverses polarity, as shown in Figure 5-6(c), then the two sources subtract from each other. Now fewer electrons are injected into the base, and consequently fewer reach the collector. Therefore, we can see that the forward bias is varied at a signal rate, producing collector current at a signal rate. The signal voltage is considerably smaller than the dc bias because we wish to maintain forward bias at all times even when the two sources subtract.

Usually in circuits in which ac and dc appear one analyzes the circuit by asking what ac "sees" and then what dc "sees." In Figure 5-6(a) we know what dc sees because we just finished this analysis in the preceding paragraph. We did add the resistance of the transformer secondary winding but it contributes very little resistance in comparison with R_B. On the other hand, ac sees R_B and the base-to-emitter resistance. This means that the ac voltage e_i will be dropped across R_B and the base-to-emitter junction. Since the primary purpose of the total circuit is to amplify weak signals, it would be advantageous not to "lose" any ac voltage. This is precisely what is happening. R_B, which is necessary for dc conditions, is accepting some of the ac voltage when indeed we would like to get as much as possible *to* the transistor. To get around this dividing of input voltage we can place a large capacitor across R_B whose reactance will be small relative to R_B. Then all that would remain would be the ac resistance offered by the forward bias junction. Our previous rough rule of thumb is appropriate here. We choose a capacitor whose reactance (at the frequency we are amplifying) is no more than one tenth of R_B.

We already know something about the resistance offered by a forward-biased diode from Chapter 4. This facilitates finding the ac equivalent circuit of the input side of the CE amplifier. We recall that this resistance is the sum of r_B and r_j. Since we operate near or below the knee in transistors, the resistance is primarily r_j and, just as in Chapter 4,

$$r_j = \frac{25 \text{ mV}}{I \text{ mA}} \tag{5-8}$$

Then in a CE amplifier the ac resistance is

$$r'_j = \frac{25 \text{ mV}}{I_B} \tag{5-9}$$

where I_B is the dc base current and r'_j is the ac diode resistance looking into the base. Once again, we repeat that this equation is an approximation and therefore subject to some inaccuracy.

This ac signal that enters and passes through r'_j is "worked on" by the transistor and becomes amplified by a factor of β_{ac}. The resulting equivalent circuit resembles the dc equivalent circuit. Figure 5-8(a) shows the transistor equivalency, Figure 5-8(b) shows a typical amplifier circuit, and Figure

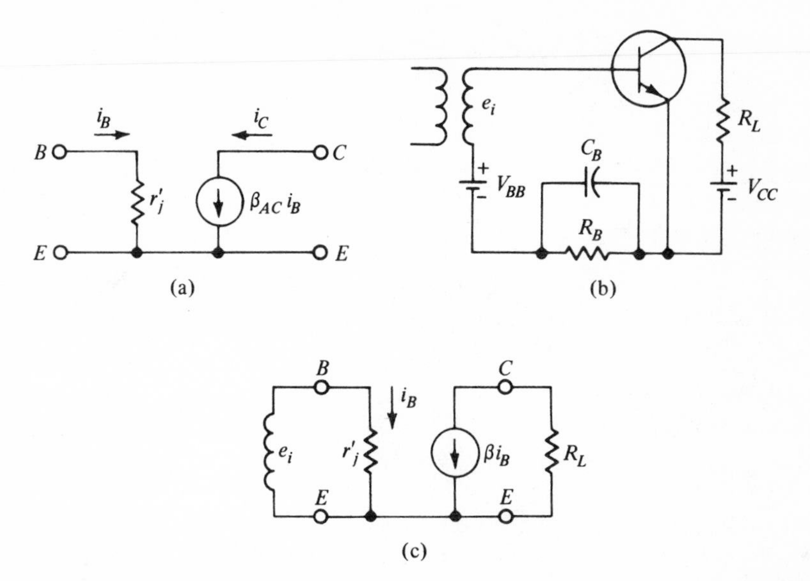

Figure 5–8 (a) Transistor ac equivalency in CE configuration. (b) CE amplifier. (c) CE amplifier ac equivalent circuit.

5-8(c) shows the resulting ac equivalent circuit that lends itself to circuit analysis. Notice that R_B does not appear in Figure 5-8(c) because C_B bypasses R_B as if it were a short circuit to ac. To solidify this information the following examples are given:

EXAMPLE 6

The signal voltage e_i that appears in Figure 5-8(b) is 10 mV peak. If the transistor has β_{ac} of 50 and operates at $I_B = 50\ \mu$A, determine (a) r'_j, (b) peak ac base current, and (c) peak ac collector current.

SOLUTION

(a)

$$r'_j = \frac{25\text{ mV}}{I_B}$$

$$= \frac{25(10^{-3})}{50(10^{-6})}$$

$$= 500\text{ ohms}$$

(b)

$$i_{B\,(\text{peak})} = \frac{V_{\text{peak}}}{R}$$

$$= \frac{10(10^{-3})}{500}$$

$$= 20\ \mu\text{A}$$

(c)

$$\begin{aligned} i_{C\,(\text{peak})} &= \beta_{ac} i_{B\,(\text{peak})} \\ &= 50(20)(10^{-6}) \\ &= 1000(10^{-6}) \\ &= 1 \text{ mA} \end{aligned}$$

EXAMPLE 7

Assume that R_B in Figure 5-8(b) is 100 kΩ. If the frequency of the signal is 1000 hertz, what value of C_B is appropriate so that its reactance "looks like" a short circuit for R_B?

SOLUTION

If we let $X_C = R_B/10$, then

$$\begin{aligned} X_C &= \frac{100 \text{ k}\Omega}{10} \\ &= 10 \text{ k}\Omega \end{aligned}$$

Since

$$X_C = 1/2\pi f C$$

then

$$\begin{aligned} C &= \frac{1}{(X_C)2\pi f} \\ &= \frac{1}{(10)(10^3)(6.28)(1)(10^3)} \\ &= 0.016\ \mu\text{F} \\ &\simeq 0.02\ \mu\text{F} \end{aligned}$$

To complete the CE amplifier we must take a look at the output circuit. Figure 5-9 shows a typical output circuit with provisions for the signal to be passed on to the next stage by way of C_C. The purpose of C_C is to block any direct current leaving V_{CC} from appearing in the next stage. The value of C_C

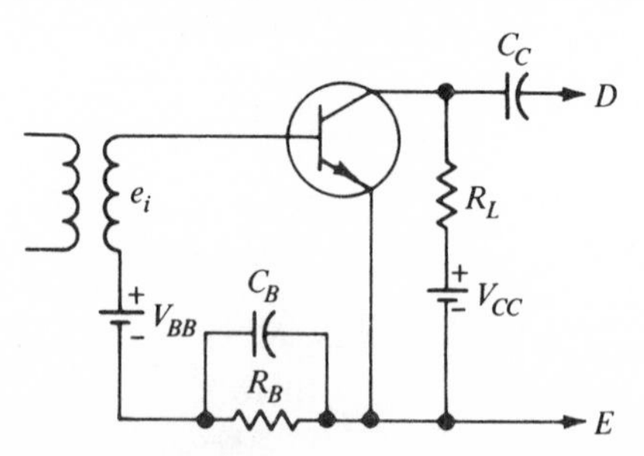

Figure 5-9 CE amplifier with signal input and output.

is such that it offers low reactance to the ac that is amplified or low reactance to the mid-frequency if a range of frequencies is amplified. More importantly, the output voltage is considered to be the voltage appearing across terminals *DE*.

If we look at the transistor side of C_C, the equation we used for dc considerations still holds true except that we must adapt it to a changing current i_C if a signal voltage appears at the input. Hence,

$$V_{CC} - i_C R_L = v_{CE} \tag{5-10}$$

Notice that if i_C increases, perhaps because of the addition of the signal voltage to the input forward bias, the drop across R_L increases, leaving less collector-to-emitter voltage. Conversely, if i_C momentarily decreases because the signal voltage is bucking the input forward bias, the drop across R_L decreases, leaving a relatively higher collector-to-emitter voltage. It is important to get a graphical picture of this behavior. Figure 5-10(a) shows what happens between the base and emitter and Figure 5-10(b) shows what simultaneously happens across the collector–emitter junction. For the period of time when no signal appears, 0 to t_1, the forward bias maintains current flow in the output and the collector–emitter voltage is a constant dc value. When ac voltage appears, the collector voltage decreases when its base–emitter voltage increases, and vice versa, because of R_L's effect, just described. When this voltage is sent through C_C, it appears as a sine wave varying above and below the zero axis, as shown in Figure 5-10(c). This means that there is no dc component present in the signal.

The most important conclusion is that a *phase inversion* in voltages occurs between input and output, that is, the signals are 180° out of phase with each other. This is one of the important characteristics that describes the CE amplifier. If a comparison is made between output voltage amplitude and input voltage amplitude, the resulting ratio yields the voltage gain A_v, which is another characteristic that helps to describe CE behavior. Mathematically,

$$A_v = \frac{v_{CE}}{v_{BE}} \tag{5-11}$$

If a change in collector voltage must equal the voltage drop across R_L, then

$$v_R = v_{CE} = i_{C\,(\max)} R_L \tag{5-12}$$

$$= \beta i_{B\,(\max)} R_L \tag{5-13}$$

Moreover,

$$v_{BE\,(\max)} = i_{B\,(\max)} r'_j \tag{5-14}$$

Substituting Equations (5-13) and (5-14) into (5-11) yields

$$A_V = \frac{\beta i_{B\,(\max)} R_L}{i_{B\,(\max)} r'_j} \tag{5-15}$$

$$= \frac{\beta R_L}{r'_j} \tag{5-16}$$

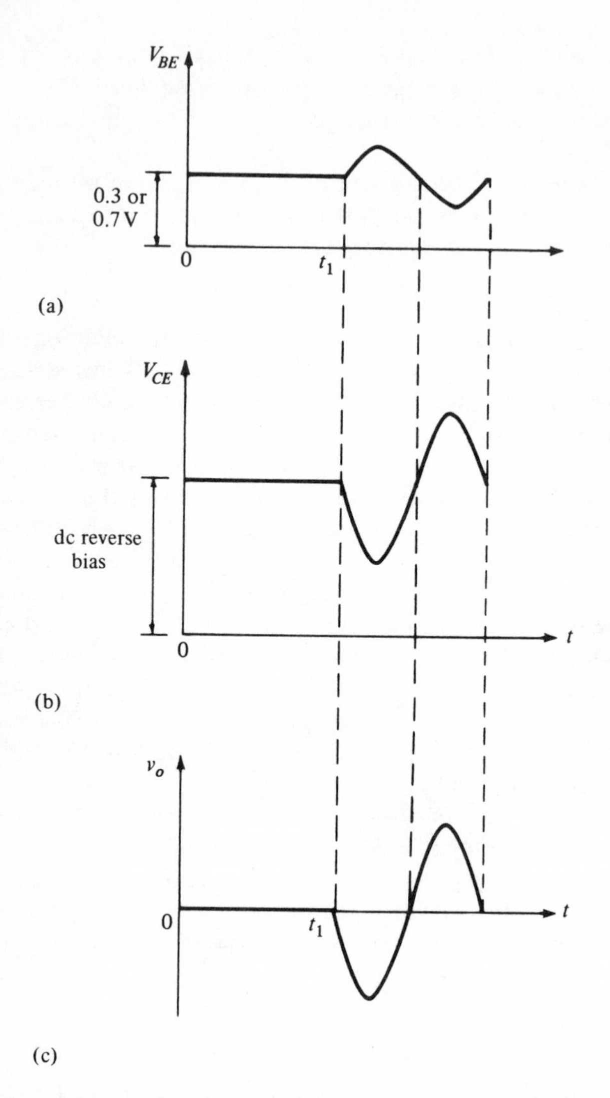

Figure 5–10 (a) Voltage between emitter and base in CE amplifier. (b) Voltage between collector and emitter. (c) Output voltage.

Since these equations must use the ac beta, we can directly indicate this by restating Equation (5-16) more appropriately as

$$A_V \simeq \frac{h_{fe}R_L}{r'_j} \tag{5-17}$$

Once again it is necessary to state that this equation will produce only an approximate answer for two reasons, (1) we know that r'_j is an approximate value for input resistance and (2) h_{fe} has a range of values depending on where the transistor operates from a dc standpoint.

EXAMPLE 8

Figure 5-11 is a silicon transistor, which has an h_{fe} of 80. V_{BB} and R_B are chosen so that $I_B = 20\ \mu A$. The resulting collector current equals

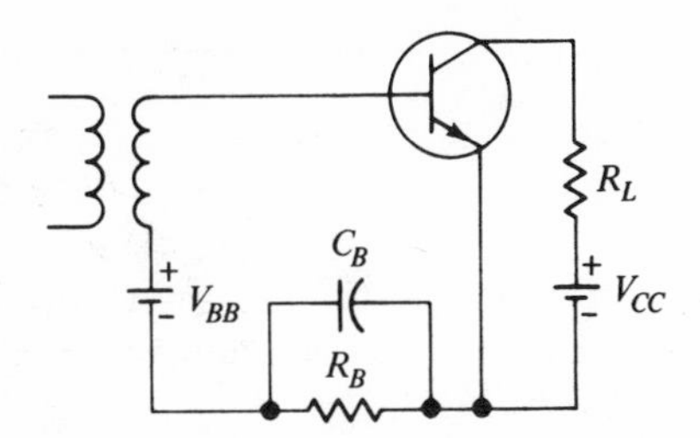

Figure 5–11 Circuit for Examples 8 and 9.

2.0 mA. If an ac voltage of 10 mV peak appears between base and emitter, determine (a) r'_j, (b) $i_{B\,(\max)}$, (c) $i_{C\,(\max)}$, (d) $i_{B\,(\max)}r'_j$, (e) $i_{C\,(\max)}R_L$, (f) the ratio of (e) to (d), and (g) A_v by using Equation (5-17).

SOLUTION

(a)

$$r'_j = \frac{25(10^{-3})}{I_B} = \frac{25(10^{-3})}{20(10^{-6})} = 1250 \text{ ohms}$$

(b)

$$i_{B\,(\max)} = \frac{e_{B\,(\max)}}{r'_j} = \frac{10(10^{-3})}{1250} = 8\ \mu A$$

(c)

$$i_{C\,(\max)} = i_{B\,(\max)}h_{fe} = 8(10^{-6})(80) = 0.64 \text{ mA}$$

(d)

$$i_{B\,(\max)}r'_j = 8(10^{-6})(1.250)(10^3) = 10 \text{ mV}$$

(e)

$$i_{C\,(\max)}R_L = (0.64)(10^{-3})(2)(10^3) = 1.28 \text{ volts}$$

(f)

$$\text{Ratio } \frac{(e)}{(d)} = \frac{1.28}{10(10^{-3})} = 128$$

(g)
$$A_V = \frac{h_{fe}R_L}{r'_j}$$
$$= \frac{80(2)(10^3)}{1.25(10^3)}$$
$$= 128$$

EXAMPLE 9

For Example 8, determine (a) the maximum and minimum values of current that the base will see and (b) maximum and minimum values of voltage that appears between base and emitter.

SOLUTION

Figure 5-10 indicates that we must add dc and ac quantities.

(a)
$$I_{B\,(\max)} = I_B + i_{B\,(\max)}$$
$$= 20(10^{-6}) + 8(10^{-6})$$
$$= 28\ \mu\text{A}$$

$$I_{B\,(\min)} = I_B - i_{B\,(\max)}$$
$$= (20 - 8)(10^{-6})$$
$$= 12\ \mu\text{A}$$

(b)
$$V_{BE\,(\max)} = 0.7 + r'_j i_{B\,(\max)}$$
$$= 0.7 + 10(10^{-3})$$
$$= 0.710 \text{ volt}$$

$$V_{BE\,(\min)} = 0.7 - 10(10^{-3})$$
$$= 0.690 \text{ volt}$$

Examples 8 and 9 demonstrate that simple Ohm's law procedures and associated circuit theory are used over and over with the addition of special rules because both ac and dc are present. Example 8 shows how an equation [part (g)] is sufficient to solve for the voltage gain as opposed to a relatively longer approach as demonstrated in parts (a) through (f).

Summarizing the overall picture shows that ac and dc are present at the input terminals, with the ac riding "piggyback." Since we can estimate the input resistance, we can obtain the ac voltage drop. As it passes through the transistor, the current is multiplied by a factor h_{fe}. Since we know that dc and ac are present at the output, we can separate the two. A ratio of the ac output voltage to the ac input voltage is called voltage gain. This gain can be approximated mathematically by the equation $A_v = h_{fe}R_L/r'_j$.

A few words are in order relative to this equation. As it stands now, we must know the h_{fe}, or β, of the transistor to predict an approximate gain.

Most manufacturers quote a range. It would be desirable to simplify this expression further, as follows. Since

$$A_v = \frac{h_{fe} R_L}{r'_j} \tag{5-18}$$

then

$$A_v = \frac{R_L}{r'_j/h_{fe}} \tag{5-19}$$

Substituting for r'_j,

$$A_v = \frac{R_L}{25\ \text{mV}/I_B h_{fe}} \tag{5-20}$$

If we assume $h_{fe} \simeq h_{FE}$, then

$$A_v = \frac{R_L}{25\ \text{mV}/I_C} \tag{5-21}$$

therefore,

$$A_v = \frac{R_L}{r_j} \tag{5-22}$$

where

$$r_j = \frac{25\ \text{mV}}{I_C} \tag{5-23}$$

or

$$r_j = \frac{25\ \text{mV}}{I_E} \tag{5-24}$$

because I_C approximately equals I_E. Recall that we already used Equation (5-24) in estimating diode resistance in a previous chapter.

Equation (5-22) is quite simple. The voltage gain is a ratio of two resistances: the first, the load resistance, is in the numerator; the second, which is in the denominator, is an estimated resistance that depends on the collector current flowing in the transistor. The source of error exists in the estimate of r_j because the 25 mV is a derived theoretical value for an ideal diode. As mentioned previously, this number ranges from 25 to 50 in practical cases. Therefore, to improve our estimate of the voltage gain, we will avoid the low-end estimate of r_j and henceforth use 30 mV instead of 25 mV. Therefore,

$$r_j \simeq \frac{30\ \text{mV}}{I_C} \tag{5-25}$$

Hopefully, this will produce more reliable answers for A_v.

EXAMPLE 10

A transistor's CE characteristics, as seen in a transistor manual, has a range of 0 to 6 mA of collector current. If a load resistance of 2 kΩ and an I_C of 3 mA are arbitrarily chosen, determine (a) voltage gain expected, (b) dc voltage drop across R_L, (c) V_{CC} required if V_{CE} is arbitrarily chosen to be equal to V_{RL}.

SOLUTION

(a)

$$r_j = \frac{30 \text{ mV}}{3(10^{-3})}$$
$$= 10 \text{ ohms}$$

$$A_v = \frac{R_L}{r_j}$$
$$= \frac{2(10^3)}{10}$$
$$= 200$$

(b)

$$V_{RL} = I_C R_L$$
$$= 3(10^{-3})(2)(10^3)$$
$$= 6 \text{ volts}$$

(c)

$$V_{CC} = I_C R_L + V_{CE}$$
$$= 6 + 6$$
$$= 12 \text{ volts}$$

EXAMPLE 11

A transistor is chosen to operate at 0.5 mA. If the voltage gain is limited to 60, determine the approximate load resistance needed.

SOLUTION

$$r_j = \frac{30 \text{ mV}}{0.5(10^{-3})}$$
$$= 60 \text{ ohms}$$

$$A_v = \frac{R_L}{r_j}$$

$$R_L = A_v r_j$$
$$= (60)(60)$$
$$= 3600 \text{ ohms}$$

5-3 Common-Base Amplifier

Regardless of the amplifier configuration that may be used, the same principles apply; that is, the dc and then ac requirements must be satisfied. If the preceding CE discussion was understood, then no problem should be encountered comprehending CB considerations. We will look first at dc and then at ac conditions, concluding with some approximate relationships.

Figure 5-12(a) shows a typical *n-p-n* transistor in a CB configuration with the necessary biases. An examination of the dc input factors results in Figure 5-12(b), where diode *D* is an ideal diode (zero ohms in forward bias) and V_{EB} is the required forward voltage (0.3 or 0.7) that must be overcome before conduction can occur. A summation of voltages in this figure produces

$$-V_{EE} + V_{BE} + I_E R_E = 0 \qquad (5\text{-}26)$$

or

$$R_E = \frac{V_{EE} - V_{BE}}{I_E} \qquad (5\text{-}27)$$

This equation permits us to calculate the resistance necessary to limit the emitter current. If V_{EE} is much larger than V_{EB} (our rules say a factor of ten is necessary), then we can approximate a little further and say that Equation (5-27) becomes

$$R_E \simeq \frac{V_{EE}}{I_E} \qquad (5\text{-}28)$$

This expression is represented by Figure 5-12(c).

If we look at the output side of Figure 5-12(a), the following equation is obtained:

$$-V_{CC} + I_C R_C + V_{CB} = 0 \qquad (5\text{-}29)$$

or

$$V_{CC} = I_C R_L + V_{CB}$$

This tells us that, just as in the CE circuit, the sum of the dc voltage drop

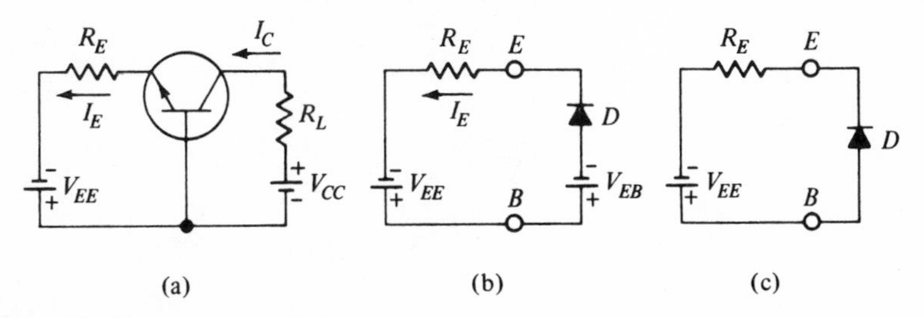

Figure 5–12 (a) CB amplifier. (b) Emitter–base dc equivalent circuit. (c) Emitter–base simplified circuit.

across R_L plus the collector–base junction must equal the reverse battery supply voltage. Figure 5-13(a) shows the output circuit and 5-13(b) shows the equivalent circuit. If we wish to approximate a little further, Figure 5-13(c) is appropriate because α is almost equal to 1.

Summarizing the dc equivalency of a CB amplifier circuit, Figure 5-14(a) shows the circuit, Figure 5-14(b) the transistor equivalency, and Figure 5-14(c) the total circuit equivalency. This last figure lends itself to circuit-theory analysis, particularly if other resistors are attached. This situation occurs when a complete amplifier is studied.

The following problem will serve as an example for biasing a CB amplifier.

EXAMPLE 12

In Figure 5-14(a), I_E is chosen to be 2.0 mA. If $V_{EE} = 12.0$ volts, $V_{CC} = 12.0$ volts, and $R_L = 5.0\ \text{k}\Omega$, determine (a) R_E, (b) I_C, (c) $I_C R_L$, and (d) V_{CB}.

SOLUTION

(a)
$$R_E \simeq \frac{V_{EE}}{I_E} = \frac{12}{2(10^{-3})} = 6\ \text{k}\Omega$$

(b)
$$I_C \simeq I_E \simeq 2\ \text{mA}$$

(c)
$$I_C R_L = 2.0(10^{-3})(5.0)(10^3) = 10 \text{ volts}$$

(d)
$$V_{CB} = V_{CC} - I_C R_L = 12 - 10 = 2.0 \text{ volts}$$

Now we can repeat the procedure established in the CE amplifier paragraphs to determine the CB amplifier ac equivalent circuit. The transformer

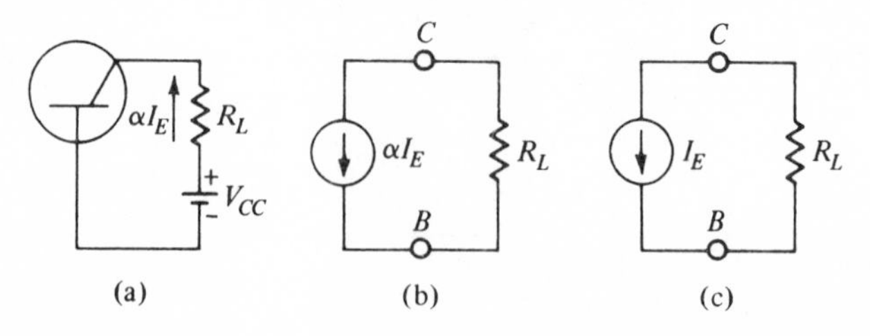

Figure 5-13 (a) CB output circuit. (b) DC equivalent circuit. (c) Simplified equivalent circuit.

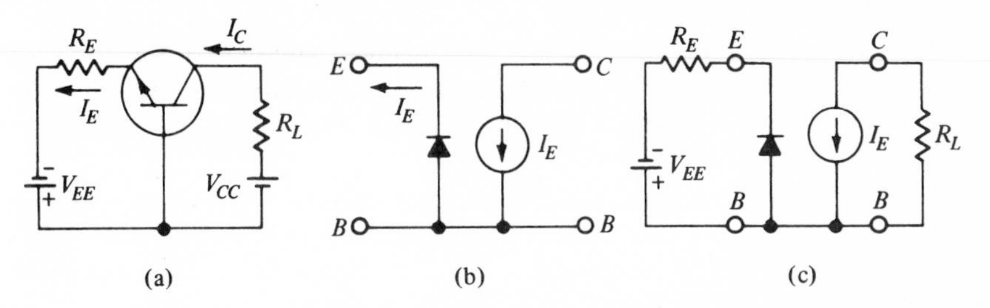

Figure 5–14 (a) CB amplifier. (b) CB transistor dc equivalent circuit. (c) CB amplifier dc equivalent circuit.

in Figure 5-15(a) is used to inject the signal into the input circuit. C_E has low reactance to ac (X_{CE} is less than 1/10 of R_E); thus most of the input signal appears across the emitter-to-base resistance. This resistance value we know to be

$$r_j = \frac{30\text{ mV}}{I_E}$$

On the output side, the output current change equals the input current change because $i_C \simeq i_E$. Therefore, the output equation holds true under ac conditions; that is,

$$v_{CB} = V_{CC} - i_C R_L \tag{5-30}$$

Figure 5-15(b) shows the transistor ac equivalence and Figure 5-15(c) shows the total circuit equivalence.

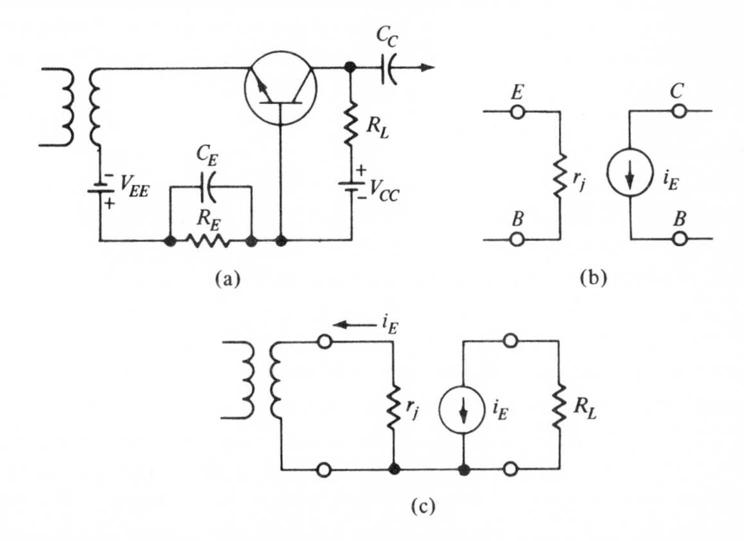

Figure 5–15 (a) CB amplifier with ac input and output. (b) Transistor ac equivalence. (c) CB amplifier ac equivalent circuit.

It must be apparent that the CE equivalent circuit looks like the CB equivalent circuit. The differences, of course, are the currents and the value of the input resistance. It is important to notice that for CB the transistor input resistance is

$$r_j = \frac{30 \text{ mV}}{I_E} \tag{5-31}$$

whereas the input resistance for a CE configuration is

$$r'_j \simeq \frac{30 \text{ mV}}{I_B} \tag{5-32}$$

or

$$r'_j = \frac{30 \text{ mV}}{I_C/h_{FE}} \tag{5-33}$$

or

$$r'_j = \frac{30 \text{ mV}}{I_E/h_{FE}} \tag{5-34}$$

assuming $I_C \simeq I_E$, or

$$r'_j = h_{FE} \frac{30 \text{ mV}}{I_E} \tag{5-35}$$

Comparing Equation (5-35) to (5-31) tells us that the CE input resistance is h_{FE} times as great as the CB input resistance. This is an important discriminating feature between CE and CB amplifiers. In addition, it permits a quick determination of input resistance if we go from one to another. To remember whether to divide or multiply r_j by h_{FE}, one has to recall that in CE amplifiers the input current I_B is much smaller than I_E in CB amplifiers. Therefore, a smaller current implies a larger resistance, which in turn implies that the CE input resistance is greater than that of CB by a factor of h_{FE}.

Finally, let us investigate the phase relationship between input and output voltages. Figure 5-16(a) shows a CB amplifier. If we assume that no signal is present, then the total forward bias is either 0.3 or 0.7 volt depending on the type of transistor used. In either case, current will flow into the collector circuit. Here a voltage drop across R_L occurs, leaving some $(V_{CC} - I_C R_L)$ for the collector-to-base junctions. This is indicated for period 0 to t_1 in Figures 5-16(b), (c), and (d). If point A in Figure 5-16(a) becomes positive momentarily, as shown in Figure 5-16(b), then the net forward bias is decreased, because the two voltages (ac and dc) are bucking. If the bias is decreased, the forward current decreases; if the forward current decreases, then the reverse current I_C decreases. A decrease in I_C reduces the voltage drop across R_L and this, in turn, permits the collector-to-base voltage to rise. Therefore, a positive-going signal (e_i) produced a positive going output (V_{CB}). If we repeat the process for a negative-going signal (A negative with

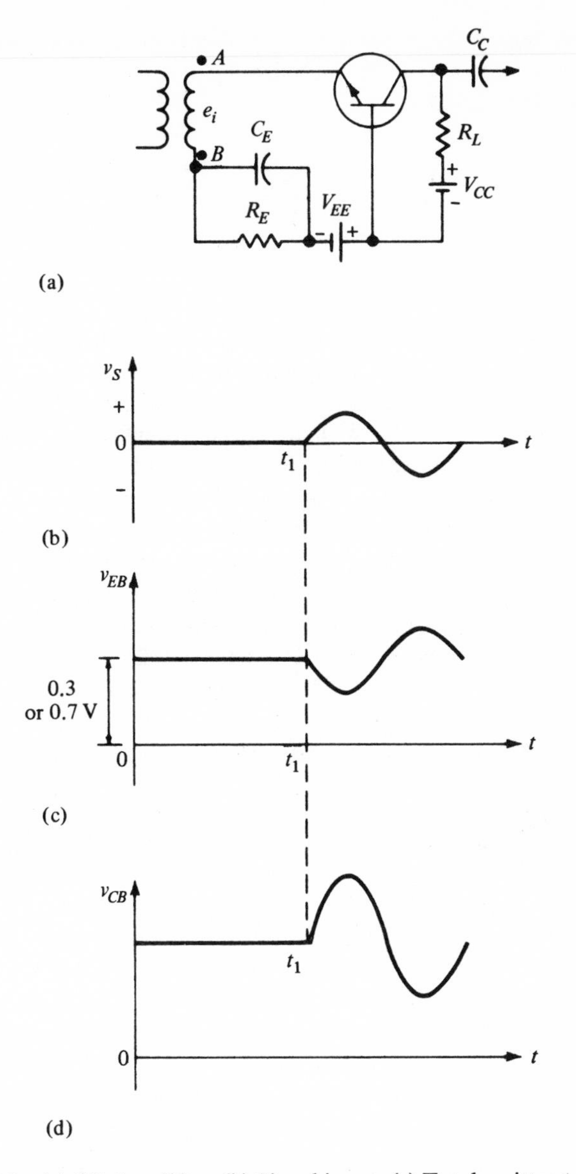

Figure 5–16 (a) CB Amplifier. (b) Signal input. (c) Total emitter-to-base voltage. (d) Total collector-to-base voltage.

respect to *B*), the forward bias is increased. This produces more collector current and a larger drop in R_L. Consequently, V_{CB} decreases. Therefore, a negative-going signal voltage (with respect to base) produces a negative-going output voltage. The conclusion we reach by looking at Figures 5-16(b) and (d) is that *no phase inversion* occurs in a CB amplifier.

The voltage gain is a comparison of the magnitudes of output to input; it is identical to that of the CE amplifier. Hence, the voltage gain in a CB amplifier is approximated by

$$A_v = \frac{R_L}{r_j} \tag{5-36}$$

where

$$r_j = \frac{30 \text{ mV}}{I_E} \tag{5-37}$$

This conclusion of equal voltage gains is a little misleading, and requires further explanation. The gains that are obtained are strictly gains for the transistor with a single load resistance. In a practical case, the load resistance has other resistors in parallel with it. In addition, one does not usually measure the gain of a transistor but the gain of the stage. Essentially, this means a comparison between a truly useful load voltage and the voltage delivered by a generator. Perhaps a simple illustration can explain this point better. Figure 5-17(a) is a CB amplifier with R_S representing the source or generator resistance. Figure 5-17(c) is the same transistor in a CE arrangement with the same generator resistance, equal to 100 ohms, for both connections. Let us assume that the h_{FE} equals 100 and I_E equals 1.0 mA. If the input equivalent circuits are drawn, Figure 5-17(b) shows an emitter-to-base resistance of 30 ohms and Figure 5-17(d) shows this resistance to be (30 × 100) = 3000 ohms for the CE arrangement. If we calculate the

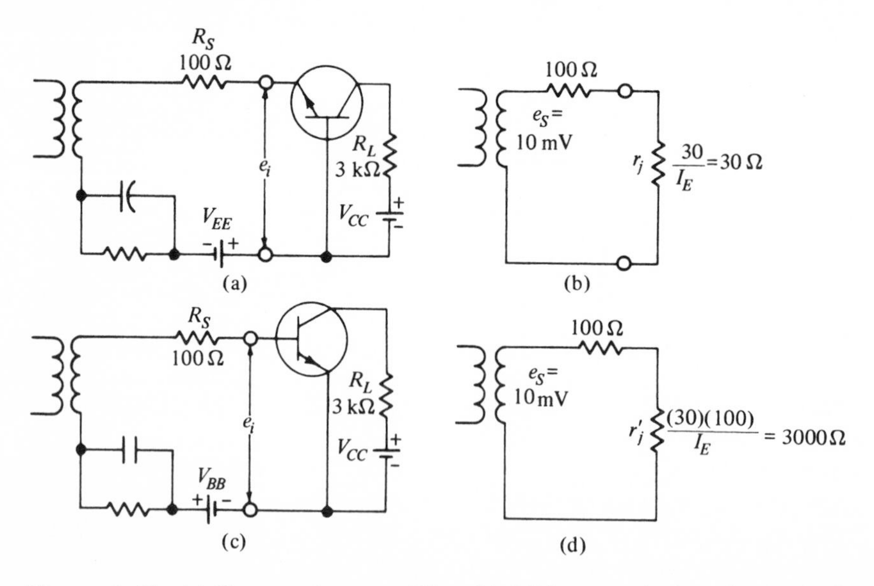

Figure 5–17 (a) Common-base amplifier. (b) AC input equivalent circuit. (c) CE amplifier. (d) AC input equivalent circuit.

transistor gain mathematically, we obtain

$$A_v = \frac{R_L}{r_j}$$
$$= \frac{3000}{30}$$
$$= 100$$

This is, of course, the gain for either arrangement. Now suppose we back up a little and assume that a 10-mV signal appears at the transformer (generator) and we wish to compare it to the voltages that appear across the 3-kΩ loads. We know that the overall voltage gain is

$$A_v = \frac{e_o}{e_i}$$

then the output voltage is

$$e_o = Ae_i$$

We know that A_V is (100); all we have to do is find e_i for both cases. In Figure 5-17(b), by voltage division,

$$e_i = 10 \text{ mV} \frac{30}{100 + 30}$$
$$= 2.3 \text{ mV}$$

Therefore, for the CB case,

$$e_o = Ae_i$$
$$= 100(2.3)(10^{-3})$$
$$= 0.23 \text{ volt}$$

For the CE case, in Figure 5-17(d),

$$e_i = 10 \text{ mV} \frac{3000}{3000 + 100}$$
$$= 9.7 \text{ mV}$$

Therefore,

$$e_o = Ae_i$$
$$= 100(9.7)(10^{-3})$$
$$= 0.97 \text{ volt}$$

Now if we wish to observe the *stage* gain, we should compare e_o to generator output.

For the CB arrangement,

$$A_v = \frac{e_o}{e_i}$$
$$= \frac{0.23}{10(10^{-3})}$$
$$= 23$$

For the CE arrangement,

$$A_v = \frac{e_o}{e_i}$$
$$= \frac{0.97}{10(10^{-3})}$$
$$= 97$$

The foregoing discussion proves that the simple transistor gain for CB and CE amplifiers are equal but that when stage gains are considered we can expect different values. The drop in overall gain for the CB case is caused by the low input resistance that the CB configuration offers. This is another discriminating feature of CB amplifiers.

5-4 Common-Collector Amplifier

The preceding section made several conclusions about input resistance. This point leads us directly to the most significant characteristic of a CC amplifier. Its input resistance is considerably larger than that of either of the other two configurations. But before we tackle this feature let us proceed through the established approach of first satisfying dc conditions and then arranging for and analyzing ac conditions.

Figure 5-18(a) shows a CC circuit, where we anticipate R_B will determine the forward bias current. Notice that C_B bypasses R_B, and thus R_B will not be part of the ac circuit. Figure 5-18(b) shows the dc path. If we make a

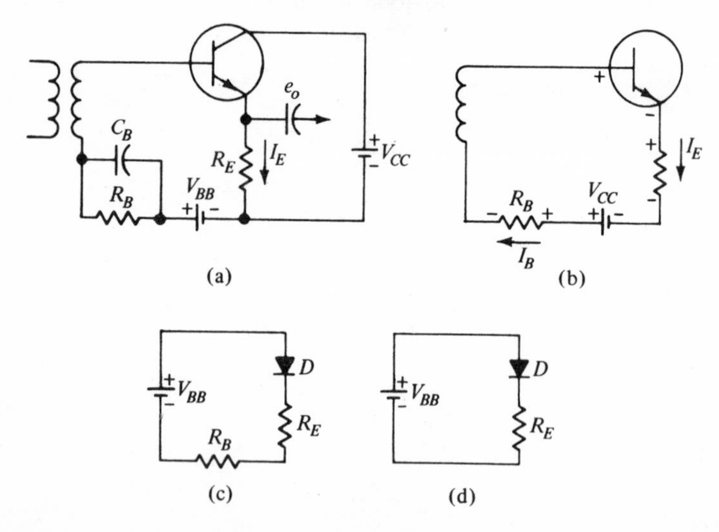

Figure 5–18 (a) CC amplifier. (b) Base–emitter dc circuit. (c) DC equivalent circuit. (d) Simplified dc equivalent circuit.

loop equation,

$$-V_{BB} + I_B R_B + V_{BE} + I_E R_E = 0 \tag{5-38}$$

then

$$I_E R_E + I_B R_B = V_{BB} - V_{BE} \tag{5-39}$$

since

$$I_B = \frac{I_C}{h_{FE}} \tag{5-40}$$

or

$$I_B \simeq \frac{I_E}{h_{FE}} \tag{5-41}$$

Then, substituting into Equation (5-39), we obtain

$$I_E R_E + \frac{I_E R_B}{h_{FE}} = V_{BB} - V_{BE} \tag{5-42}$$

$$I_E \left(R_E + \frac{R_B}{h_{FE}} \right) = V_{BB} - V_{BE} \tag{5-43}$$

or

$$I_E = \frac{V_{BB} - V_{BE}}{R_E + (R_B/h_{FE})} \tag{5-44}$$

This equation can be simplified as follows:

If V_{BB} were much greater than V_{BE} (by a factor of ten), the equivalent circuit would look like Figure 5-18(c). Further, if R_B/h_{FE} were much smaller than R_E (that is, $R_E/10$), the resulting equivalent circuit would look like Figure 5-18(d) and Equation (5-44) would simplify to

$$I_E \simeq \frac{V_{BB}}{R_E} \tag{5-45}$$

Once again, caution must be taken whenever approximations such as these are used.

Looking at the output side of Figure 5-18(a) and making a loop equation produces

$$V_{CC} = V_{CE} + I_E R_E$$

This simply states that the sum of the drops of V_{CE} and $I_E R_E$ must equal to the source voltage V_{CC}. One normally expects to operate the transistor approximately in the middle of the characteristic curves. Therefore, V_{CE} in the preceding equation should not be near either extreme of the family of curves. Let us summarize dc conditions with two examples.

EXAMPLE 13

In Figure 5-18(a), $R_E = 10\ \text{k}\Omega$, $R_B = 24\ \text{k}\Omega$, $V_{BB} = 12$ volts. The germanium transistor has an h_{FE} of 80. Determine I_E.

SOLUTION

$$\frac{R_B}{h_{FE}} = \frac{24\ \text{k}\Omega}{80}$$
$$= 300\ \text{ohms}$$

Therefore,

$$\frac{R_B}{h_{FE}} \ll R_E$$

and

$$V_{BE} \ll V_{BB}$$

Thus, using Equation (5-45)

$$I_E = \frac{V_{BB}}{R_E}$$
$$= \frac{12}{10\ \text{k}\Omega}$$
$$= 1.2\ \text{mA}$$

EXAMPLE 14

Suppose that R_E in Example 13 were changed to $1.0\ \text{k}\Omega$ to satisfy an ac condition. How would the dc condition change?

SOLUTION

Since R_B/h_{FE} is not much smaller than R_E, this factor must be included in the calculation for I_E; and since $R_B/h_{FE} = 300$ ohms,

$$I_E = \frac{V_{BB}}{R_E + (R_B/h_{FE})}$$
$$= \frac{12}{(1.0 + 0.3)(10^3)}$$
$$= 9.2\ \text{mA}$$

Now let us investigate the phase relationship between output and input signals. Figure 5-18(a) is repeated in Figure 5-19(a). With no signal present at the transformer from zero to time t_1, ordinary forward bias appears between the base and emitter as shown in Figure 5-19(c). At the same time, direct current flows in the emitter circuit, producing a drop equal to $I_E R_E$.

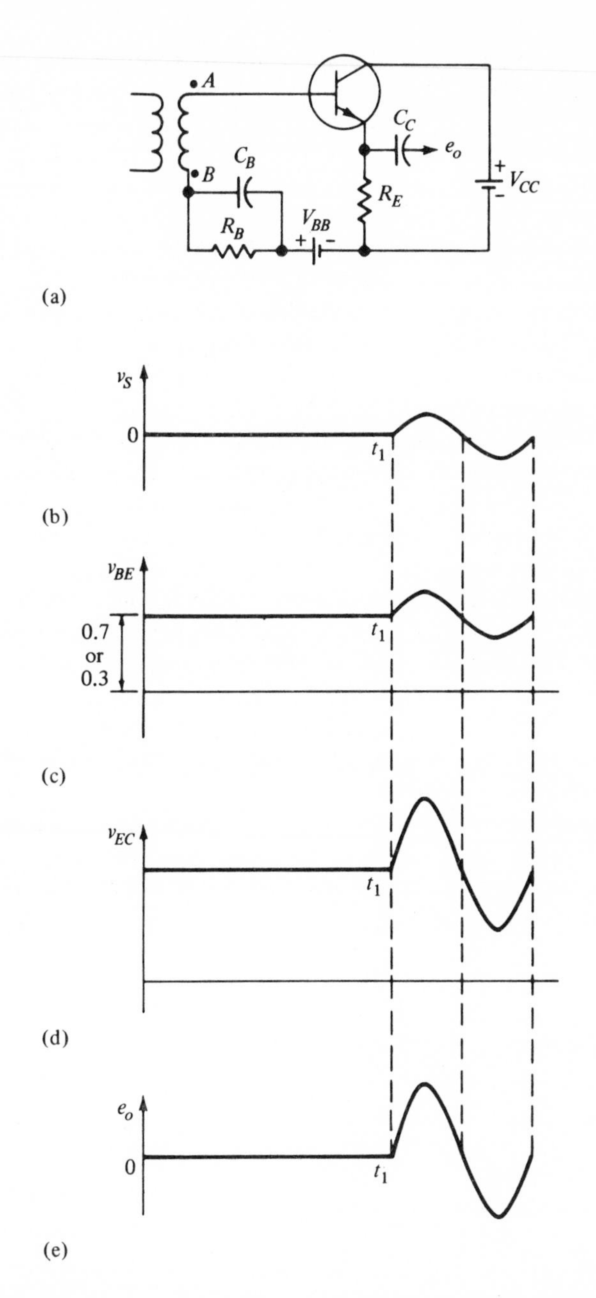

Figure 5–19 (a) CC amplifier with ac input and output. (b) Signal input. (c) Total base-to-emitter voltage. (d) Total emitter-to-collector voltage. (e) Output voltage.

After t_1, when an ac signal appears at the transformer and point A momentarily becomes more positive than point B, the forward bias increases because this voltage aids the dc bias. More forward bias causes a larger current to flow in the emitter circuit. This produces a relatively larger voltage between emitter and collector, as shown in Figure 5-19(d). Conversely, when A becomes more negative than B, the voltages buck each other. This produces less current flow in the emitter resistor, and thus the voltage decreases. When this voltage passes through capacitor C_C, the dc component is eliminated as shown in Figure 5-19(e). The net relationship between input and output voltage can be seen in Figures 5-19(b) and (e); it can be seen that no phase inversion occurs in a CC amplifier. Therefore, the only configuration that produces a phase inversion is the CE amplifier.

Now if we wish to investigate what kind of voltage gain is obtainable from a CC circuit it is worth our while to resort to an ac equivalent circuit. Figure 5-19(a) is repeated in Figure 5-20(a). In Figure 5-20(b) we see that the input signal is divided between r'_j and R_E. This division, of course, leaves that much less voltage for the transistor to "work on." This effect is that of neglecting to put capacitor C_B across R_B. Reviewing briefly, R_B is in the circuit to ensure proper dc conditions. Since it also affects the ac conditions we placed a low-reactance capacitor across it. The question may arise, why can't we do the same for R_E? The answer is that we must leave R_E unbypassed because this is the resistor across which we wish to develop an ac voltage output, which in turn will be applied to the next stage. Hence we must live with R_E affecting our input circuit.

Returning to the ac equivalent circuit, we notice that the transformer sees r_j and R_E in series. We cannot add these two resistances because they have unequal currents passing through them. However, we can find this effective resistance in the following manner:

$$R_i = \frac{e_i}{i_B} \tag{5-46}$$

$$= \frac{i_B r'_j + i_E R_E}{i_B} \tag{5-47}$$

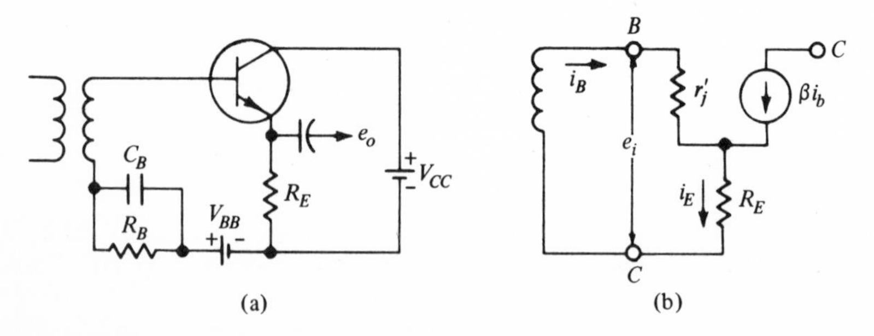

Figure 5–20 (a) CC amplifier. (b) AC equivalent circuit.

Since $r_j' = \beta r_j$ and $i_E \simeq \beta i_B$,

$$R_i = \frac{i_B \beta r_j + i_B \beta R_E}{i_B} \tag{5-48}$$

$$R_i = \beta(r_j + R_E) \tag{5-49}$$

Equation (5-49) tells us that R_i for a CC amplifier is larger than for a CB or CE amplifier because R_E is included. In some instances, this higher resistance is beneficial and therefore a CC amplifier is employed.

To find the voltage gain we must compare output voltage to input.

If

$$e_o = i_E R_E \tag{5-50}$$

$$= \beta i_B R_E \tag{5-51}$$

and

$$e_i = i_B R_i \tag{5-52}$$

then

$$A_v = \frac{e_o}{e_i} = \frac{\beta i_B R_E}{i_B \beta(r_j + R_E)} \tag{5-53}$$

or

$$A_v = \frac{R_E}{r_j + R_E} \tag{5-54}$$

where $r_j = 30\ \text{mV}/I_E$.

Once again we extracted a voltage-gain equation where only resistances are employed. A close look at Equation (5-54) tells us that the voltage gain will always be less than one. If R_E is much greater than r_j (a factor of 10 or more), then the Equation (5-54) produces a gain near one, since

$$A_v \simeq \frac{R_E}{R_E} \tag{5-55}$$

$$= 1$$

The fact that a voltage gain of less than one is achieved does not rule out the use of a CC amplifier. It should be recalled that it has a higher input resistance than either of the other two configurations. In some instances this is handy. Let us look at an example to prove the point.

As a preface it is necessary to explain about characteristic internal resistance. Nearly every electrically active device has an equivalent circuit of a voltage source in series with a resistance or a current source with a resistor in parallel. It all depends on how the device behaves. As an example, a battery with its internal resistance can be portrayed as an EMF in series with its internal resistance. On the other hand, a transistor is usually portrayed as a current generator ($h_{fe} i_B$) in parallel with a resistance ($1/h_{oe}$). For our example, we can "equivalentize" a microphone as a voltage source in series with its characteristic resistance.

EXAMPLE 15

A microphone by nature generates a voltage when audio energy is fed to it. If a microphone produces 0.5 mV, determine how much voltage will be developed across the resistance that a CE amplifier and a CC amplifier offer at their respective input terminals. For simplicity, assume that r_j for both configurations is 30 ohms, R_E is 3.0 kΩ, h_{fe} is 100, and the internal resistance of the microphone is 50 kΩ.

SOLUTION

Figure 5-21(a) is the simple equivalent circuit of the microphone and input to the CE amplifier and Figure 5-21(b) shows the CC equivalency.
(a) For the CE case,

$$\begin{aligned} \beta r_j &= (100)(30) \\ &= 3000 \text{ ohms} \end{aligned}$$

The voltage input to CE is a proportional part of total voltage,

$$\begin{aligned} \text{V input to CE} &= 0.5 \text{ mV} \frac{3000}{50{,}000 + 3000} \\ &= 0.0283 \text{ mV} \end{aligned}$$

(b) For the CC case,

$$\begin{aligned} \beta(r_j + R_E) &= 100(30 + 2000) \\ &= 203 \text{ k}\Omega \end{aligned}$$

$$\begin{aligned} \text{V input to CC} &= 0.3 \text{ mV} \frac{203(10^3)}{(203 + 50)(10^3)} \\ &= 0.4 \text{ mV} \end{aligned}$$

Notice that the CC circuit managed to receive a greater portion of the microphone voltage. This is quite important because, in any system, it is not desirable to "lose" the voltage across the internal impedance. (A dry cell becomes worthless when the internal impedance is much greater than the

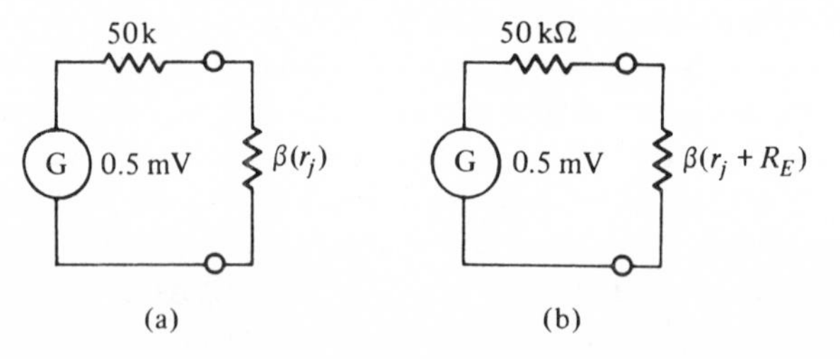

Figure 5–21 Equivalent circuit for high-impedance microphone feeding. (a) CE configuration. (b) CC configuration.

load.) Despite the fact that the CC stage will not amplify this voltage we have retained most of the microphone output and now we can pass it on to a CE stage where we can expect to boost the amplitude considerably.

EXAMPLE 16

A CC stage uses a germanium transistor whose h_{FE} is 60. If $R_E = 5.0\ \text{k}\Omega$, $R_B = 12.0\ \text{k}\Omega$, $V_{BB} = 6$ volts, $V_{CC} = 12$ volts, for Figure 5-20(a), determine (a) I_E, (b) r_j, (c) R_i, and (d) A_v.

SOLUTION

(a)
$$\frac{R_B}{h_{FE}} = \frac{12\ \text{k}\Omega}{60}$$
$$= 200 \text{ ohms}$$

Since $R_B/h_{FE} \ll R_E$ and $V_{BE} \ll V_{BB}$,

$$I_E \simeq \frac{V_{BB}}{R_E}$$
$$= \frac{6.0}{5.0\ \text{k}\Omega}$$
$$= 1.2\ \text{mA}$$

(b)
$$r_j = \frac{30\ \text{mV}}{1.2\ \text{mA}}$$
$$= 25 \text{ ohms}$$

(c)
$$R_i = h_{FE}(r_j + R_E)$$
$$= 60(25 + 5.0\ \text{k}\Omega)$$

Since $r_j \ll R_E$,

$$R_i = 60(5.0\ \text{k}\Omega)$$
$$= 300\ \text{k}\Omega$$

(d)
$$A_v = \frac{R_E}{r_j + R_E}$$
$$= \frac{5\ \text{k}\Omega}{25 + 5.0\ \text{k}\Omega}$$

Since $r_j \ll R_E$,

$$A_v \simeq \frac{5}{5}$$
$$= 1$$

As a final note we add that the CC amplifier is also called an emitter follower (EF) because the emitter signal (output) follows the signal applied to the base in phase relationship and nearly in amplitude.

5-5 Comparing the Three Basic Configurations

Previous examples have demonstrated individual characteristics of each configuration. Where an individual type is used depends on the demands of the design. In any event, it is necessary to generalize and memorize these descriptive characteristics. Table 5-1 is essentially a summary of formulas for the three configurations. Employing these equations would give approximate answers to simple amplifier arrangements.

Table 5-1

	CB	*CE*	*CC*
R_i	r_j	βr_j	$\beta(r_j + R_E)$
A_v	$\dfrac{R_L}{r_j}$	$\dfrac{R_L}{r_j}$	$\dfrac{R_E}{R_E + r_j}$
Phase shift (from input to output)	0°	180°	0°
A_i	α	β	β

To describe these configurations completely it would help to add another column—namely, R_o, the resistance that the transistor offers to the rest of the circuit on the output side. It is placed directly across the current generator in the equivalent circuit. An approximate value is $1/h_{oe}$ for CE and CC, and $1/h_{ob}$ for CB. These values are obtainable from the family of curves and, if you recall, we said that R_o is a relatively large number in comparison to typical load resistances and therefore we ignored R_o. We find it necessary to mention R_o because all circuits are analyzed with the view of the effective resistance that ac sees looking from these terminals or those terminals. A classic example was the microphone problem in the preceding section. The question there involved the ac resistance the microphone saw looking into CC and CE amplifiers. The question can also be asked from other locations in the circuit. As an example, suppose transistor one (T_1) feeds its signal to transistor two (T_2). At the output of T_1 one can ask, what resistance does the ac signal see as it leaves the transistor?

Figure 5-22(a) shows a two-stage CE amplifier. Notice that if C_C and all batteries have zero ohms to ac the circuit of Figure 5-22(b) results. Now the question asked is what ac resistance the current generator sees across its terminals. The reason for this question is that *this resistance* is what we use

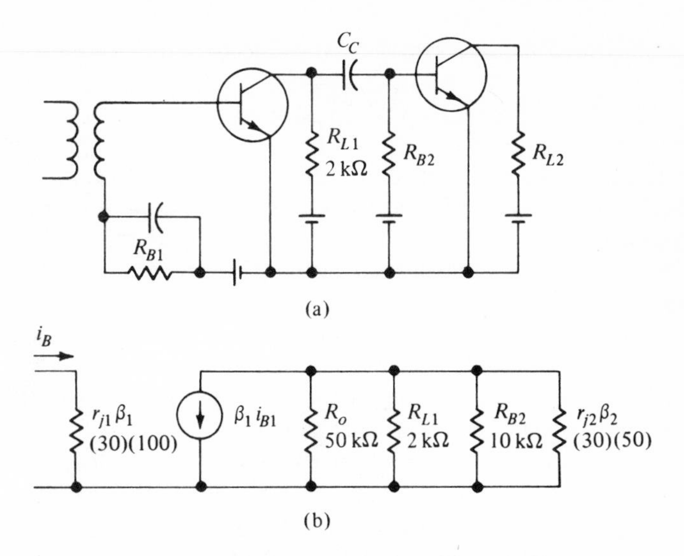

Figure 5–22 (a) Two-stage CE amplifier. (b) Equivalent circuit for stage one.

for R_L in the equation for voltage gain, where $A_v = R_L/r_j$. Up to now we used simple single-transistor amplification, but now the inclusion of all these resistances in the calculation results in stage amplification. More descriptively, stage amplification is a comparison of the voltage that appears at input terminals of *stage two* to the voltage that appears at input of terminals of *stage one*—indeed a *complete stage*. From Figure 5-22(b) it can be seen that the parallel combination of R_{L1}, R_{B2}, and $r_{j_2}\beta_2$ is much smaller than R_o, so we can ignore R_o:

$$R_{||} = \frac{1}{(1/2000) + (1/10{,}000) + (1/1500)}$$

$$= \frac{1}{1.27(10^{-3})}$$

$$= 790 \text{ ohms}$$

$$R_{||} \ll R_o$$

$$A_v = \frac{R_{||}}{r_{j1}}$$

$$= \frac{790}{30}$$

$$= 26.3$$

Notice that the voltage-gain calculation for stage one required input (bias and base-to-emitter) resistances of stage two. It must be remembered that all the resistance that the current generator sees are involved in the calculation.

If we simply used $R_L = 2\ \text{k}\Omega$, the gain would be

$$A_v = \frac{2\ \text{k}\Omega}{30\ \Omega} = 66.7$$

This answer is quite incorrect.

The foregoing example may seem like a digression, but it serves to emphasize that the expressions in Table 5-1 are restricted to simple amplifiers. However, these expressions are applicable if we redefine terms such as R_L in the example cited.

The same holds true for current gain A_i. Beta is a theoretical gain and if we compare the current entering T_2 with the current entering T_1, we would have the actual current gain. Looking at the equivalent circuit βi_B is the current generator and all its current is split among R_o, R_L, R_{B2}, and $r_{j2}\beta_2$. The practical output current (that which enters $r_{j2}\beta_2$) obviously will be some fractional part of βi_B. Therefore, the current gain will be less than β or h_{fe}. Once again, the Table 5-1 values are not appropriate unless we remember these restrictions or redefine our terms. To redefine terms, let us derive an equation for A_i. Classically, current gain is defined as a ratio of output current divided by input current.

$$A_i = \frac{I_o}{I_i} = \frac{e_o/R'_o}{e_i/R_i} = \frac{e_o}{e_i}\frac{R_i}{R'_o}$$

We know that e_o/e_i is voltage gain. Therefore,

$$A_i = A_v \frac{R_i}{R'_o}$$

where R_i is the input resistance to the transistor under consideration and R_o is the *resistance to which we ultimately wish to deliver current.* In the example problem above, R_i is $\beta_1 r_{j1}$, the input to T_1, and R'_o is $B_2 r_{j2}$, the input to T_2. Hence the current gain for stage one is

$$A_i \simeq A_v \frac{R_{i1}}{R'_o} = 26.3\,\frac{(100)(30)}{1500} = 52.6$$

Notice that this gain is considerably less than h_{fe}, which is 100. But once again, we considered the stage gain and not just the transistor gain with only one load resistance.

The beauty of our approach is that the stage voltage gain is relatively simple to find ($A_v = R_L/r_j$, where R_L has a new meaning) and then the stage current gain uses this voltage gain in its evaluation. We could have used the equivalent circuit and used a current division approach to see how much of $\beta_1 i_B$ reached $\beta_2 r_{j_2}$ and then compared it to i_B. However, this would take a little longer and the other approach seems simpler.

To prove the validity of the simple current-gain formula let us apply it to a simple amplifier where the only load resistance is R_L. Let us use Figure 5-22, where all circuit constants are given, but omit T_2 and its associated circuit.

$$A_v = \frac{R_L}{r_j} = \frac{2\text{ k}\Omega}{30\ \Omega} = 66.7$$

$$\begin{aligned} A_i &= A_v \frac{R_i}{R'_o} \\ &= 66.7 \frac{\beta_1 r_{j1}}{2\text{ k}\Omega} \\ &= \frac{66.7(100)(30)}{2000} \\ &= 100 \end{aligned}$$

This, incidentally, is h_{fe} of T_1 and precisely what we said A_i should be under simple amplifier conditions. Notice that R'_o happens to be R_L because it is the only useful resistance to which we wish to deliver current.

To complete the comparison of the three configurations, it might be worthwhile to use the expressions in Table 5-1 with actual numbers. Let us assume that the same transistor will be used in the three configurations. Let us also assume that the operating point will be $I_E = 1$ mA, and that $h_{fe} = 50$ and $R_E = R_L = 1.0$ kΩ. Table 5-2 shows the values obtained for each configuration. Its values tend to demonstrate the differences between configurations most dramatically.

Table 5-2

	CB	*CE*	*CC*
R_i	30 ohms	1500 ohms	51,500 ohms
A_i	0.99	100	100
A_v	33.3	33.3	0.97

5-6 Methods Used for dc Bias

Up to this point, dc biasing required two sources, one for the forward base-to-emitter junction and another for the reverse base-to-collector junction. Indeed, circuits can be connected with two sources for each transistor, but would it not be a better practice to use only one source for all the transistors? The student may recall that in the process of replacing batteries in a transistor radio he may have changed only one 9-volt battery or four penlite series-connected batteries. In this paragraph, single-source connections will be investigated. One arrangement will be used because of its simplicity, the other because of its stabilizing effect.

Figure 5-23(a) shows the method of dc biasing we have been using in a CE amplifier. Figure 5-23(b) shows a simple method wherein we use only one source, namely the collector supply V_{CC}. If a dc loop equation for the forward-biased EB is extracted, it yields

$$-V_{CC} + I_B R_B + V_{BE} = 0 \tag{5-56}$$

or

$$R_B = \frac{V_{CC} - V_{BE}}{I_B} \tag{5-57}$$

If $V_{CC} \gg V_{BE}$, then

$$R_B \simeq \frac{V_{CC}}{I_B} \tag{5-58}$$

Notice that capacitor C_B is needed to prevent direct current from going through the input transformer. This R_B value is usually quite large because I_B for small signal amplifiers is in the microampere range. One might rather hastily conclude that any alternating current coming from the transformer would also proceed through R_B and through the battery instead of entering the transistor. This does not happen because R_B is considerably larger than the input transistor resistance βr_j and therefore most of the current enters the transistor. As an example, R_B may be 100 kΩ and R_i may be approximately 1.0 kΩ.

Suppose a resistor is added in the emitter circuit of Figure 5-23(b), as shown in Figure 5-24. In this case the loop equation involving forward bias is

$$-V_{CC} + I_B R_B + V_{BE} + I_E R_E = 0 \tag{5-59}$$

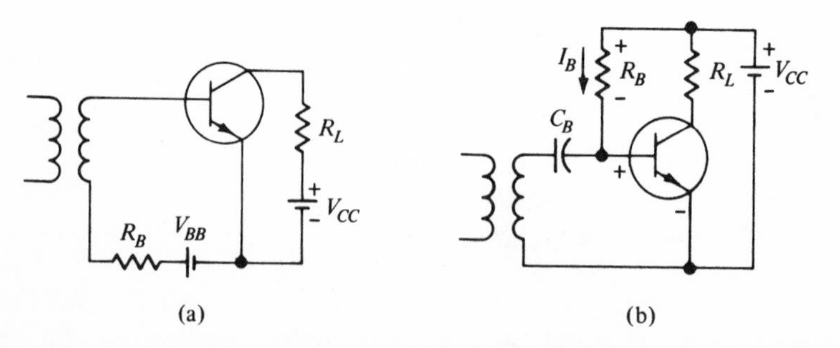

Figure 5–23 (a) Typical two-battery biasing. (b) Simple one-battery biasing.

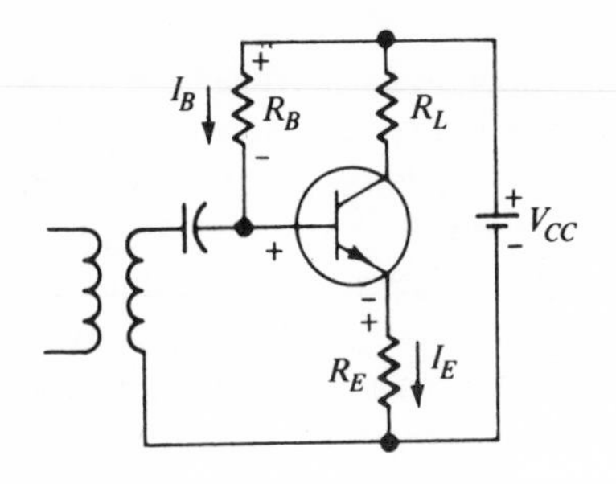

Figure 5–24 Biasing with base and emitter resistors.

If

$$I_E \simeq I_C \qquad \text{and} \qquad V_{CC} \gg V_{BE}$$

then

$$I_B R_B + I_C R_E = V_{CC} \tag{5-60}$$

or

$$R_B = \frac{V_{CC} - I_C R_E}{I_B} \tag{5-61}$$

This equation simply provides a means of determining R_B when all the other quantities are known. However, let us turn Equation (5-59) around and solve for V_{BE}. This produces

$$V_{BE} = V_{CC} - I_B R_B - I_E R_E \tag{5-62}$$

This equation tells us that the battery V_{CC} supplies the necessary forward bias when the voltages $I_B R_B$ and $I_E R_E$ are subtracted from it. Therefore, if I_E should *increase* for some reason, the bias voltage would *decrease*.

Earlier in the text it was mentioned that the collector current, which depends on the emitter current, increases when the temperature increases. An increase in temperature produces more current and so on until the transistor burns itself out. However, with an emitter resistor in the circuit, an increase of current that may be caused by a temperature rise will automatically produce some cancellation because the forward bias is decreased by the small increase in the $I_E R_E$ drop. A larger temperature rise, causing a larger leakage current, will commensurately be canceled by a larger reduction in forward bias. The net effect on the transistor is to improve its stability; that is, its operation remains relatively stable with temperature changes. To produce this kind of result, the output current I_C ($\simeq I_E$) must have had some effect on the input current I_B. This phenomenon, called feedback, is frequently used in electronics. A quick look at simple biasing, as in Figure 5-23, reveals that there is no feedback and therefore one can expect poor stability. However, if the transistor is used in which variations are tolerable, there is no reason why simple biasing cannot be used.

If better stability is desired, the circuit of Figure 5-24 is changed slightly to look like that in Figure 5-25. For the price of one more resistor the stability is further improved. There are numerous other circuits that provide a means for biasing but this is the one most frequently used.

Since this is the most practical circuit it requires some investigation. The forward bias is achieved by the summation of the voltage drops that appear between the base and emitter. With the polarities as indicated the voltage across R_2 tends to forward-bias E to B and the voltage across R_E tends to reverse-bias E to B. Obviously, the voltage across R_2 has to be greater by 0.3 or 0.7 volt, depending on the type of transistor used. Therefore, we have a relationship that says

$$V_{R_2} = V_{R_E} + 0.7 \text{ (or 0.3)} \tag{5-63}$$

Next, Figure 5-25 shows that current flows from the battery V_{CC} through R_1 and then splits, with I_B going to the transistor and the remaining portion through R_2. It is necessary to establish some criteria if we wish to be able to select values of resistances for R_1 and R_2. For reasonable stability, it is safe to assume that the current I_2 through R_2 is

$$I_2 \simeq 5I_B \tag{5-64}$$

If this is so, then

$$R_2 = \frac{V_{R_2}}{5I_B} \tag{5-65}$$

Since we have a relationship, Equation (5-63), for determining V_{R_2}, it is a simple matter to solve for R_1. Finishing up, by Ohm's law,

$$R_1 = \frac{V_{CC} - V_{R_2}}{6I_B} \tag{5-66}$$

where $6I_B$ is the sum of both transistor and bleeder current (current through R_2). For better stability a larger current than $5I_B$ can be assumed. Let us investigate an example.

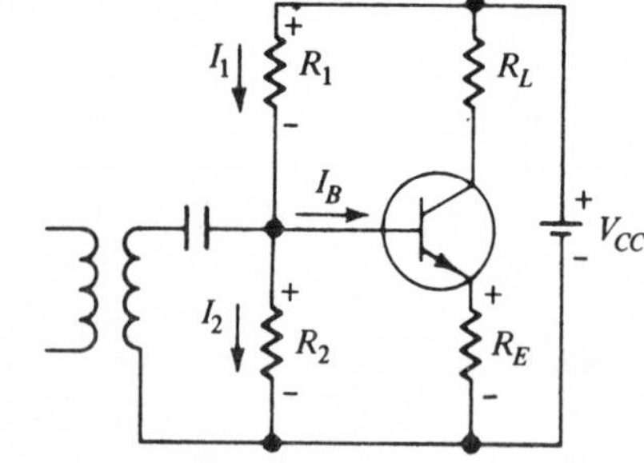

Figure 5–25 Voltage-divider emitter-resistance type of biasing.

EXAMPLE 17

A circuit as shown in Figure 5-25 is used with a silicon transistor. For proper operation an I_B of 50 μA, I_C of 2.5 mA, and R_E of 1000 ohms are selected. Determine R_1 and R_2 if $V_{CC} = 9$ volts.

SOLUTION

$$V_{R_E} = (2.5)(10^{-3})(1)(10^3)$$
$$= 2.5 \text{ volts}$$

$$V_{R_2} = V_{RE} + 0.7$$
$$= 2.5 + 0.7$$
$$= 3.2 \text{ volts}$$

Assuming that the bleeder current equals $5I_B$,

$$R_2 = \frac{V_{R_2}}{5I_B}$$
$$= \frac{3.2}{(5)(50)(10^{-6})}$$
$$= 12.8 \text{ k}\Omega$$

$$R_1 = \frac{V_{CC} - V_{R_2}}{6I_B}$$
$$= \frac{9 - 3.2}{6(50)(10^{-6})}$$
$$= 19.4 \text{ k}\Omega$$

Occasionally it is necessary to obtain information from this type of biasing when all resistances are known. As an example, the question that may be asked is, what is the approximate value of I_C if R_1, R_2, and R_E are known for a silicon transistor in Figure 5-25? One must remember that the bleeder current through R_2 is unknown because the designer's criterion is unknown. However, since the current drawn by the base is assumed to be smaller than the bleeder current, we can say

$$V_{R_2} \simeq V_{CC} \frac{R_2}{R_1 + R_2} \tag{5-67}$$

And, reversing the process,

$$V_{R_E} = V_{R_2} - 0.7 \text{ (or 0.3)} \tag{5-68}$$

Finally,

$$I_C \simeq I_E = \frac{V_{RE}}{R_E} \tag{5-69}$$

EXAMPLE 18

In Figure 5-25, $R_1 = 12\ \text{k}\Omega$, $R_2 = 6\ \text{k}\Omega$, $R_E = 0.5\ \text{k}\Omega$, and $V_{CC} =$ 12 volts. If the transistor is silicon, determine the approximate value of I_C.

SOLUTION

$$V_{R_2} \simeq \frac{V_{CC}R_2}{R_1 + R_2}$$
$$= 12\frac{6\ \text{k}\Omega}{(12 + 6)\ \text{k}\Omega}$$
$$= 4.0 \text{ volts}$$

$$V_{R_E} = V_{R_2} - 0.7$$
$$\simeq 4.0 - 0.7$$
$$= 3.3 \text{ volts}$$

$$I_C \simeq \frac{V_{R_E}}{R_E}$$
$$= \frac{3.3}{0.5(10^3)}$$
$$= 6.6 \text{ mA}$$

Usually a large capacitor is placed across R_E because it is desired not to have any ac voltage drop across it. Keep in mind R_E is placed in the circuit primarily for dc conditions and anytime, as we did previously, we wish to cancel a resistor's influence on ac we placed a low-reactance capacitor across it.

If we turn our attention to a CC amplifier to see if we can arrange single-source biasing, one should realize that a CC is similar to a CE except that the load resistor is shifted to the emitter lead from the collector. Using this approach we repeat the two methods of biasing for CC and Figure 5-26 results. For simple biasing as in Figure 5-26(a),

$$-V_{CC} + I_BR_B + V_{BE} + I_ER_E = 0 \tag{5-70}$$

Ignoring V_{BE},

$$R_B = \frac{V_{CC} - I_ER_E}{I_B} \tag{5-71}$$

In Figure 5-26(b), the same technique is used in selecting resistances as we used in the CE. One word of caution, however. In a CE circuit V_{CC} is equal to the sum of the drops across R_L, the collector-to-emitter, and R_E. In the CC circuit V_{CC} is equal to the sum of the drops across the collector-

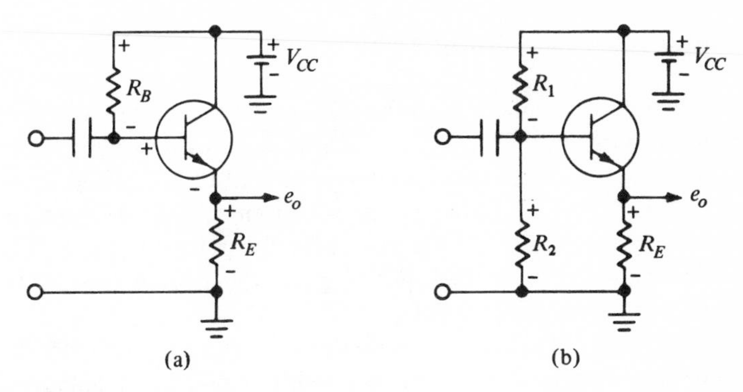

Figure 5–26 (a) CC simple biasing. (b) CC voltage-divider emitter-resistance biasing.

to-emitter and R_E. This is important when initial operating conditions are chosen. For example, if we use a 12-volt source in a CE circuit, we may choose to have 6 volts dc drop across R_L, 3 volts C to E, and 3 volts across R_E. If the same battery were used in CC circuit, the selection might be 6 volts from C to E and 6 volts across R_E. This example by no means implies that one is restricted in voltage-drop selection. The CE selection could have been 6 volts across R_E, and 3 volts each across C to E and R_L. There is indeed flexibility. One must remember only to avoid C-to-E voltages below approximately 1 volt and voltages near V_{CC}. A subsequent paragraph will illustrate this graphically.

Finally, Figure 5-27 shows the equivalent bias techniques in CB configuration. Figure 5-27(a) shows the simple bias and Figure 5-27(b) the divider-emitter bias, which yields better stability. Lack of stability is not as critical in CB as in CE circuits because temperature effects are multiplied by a factor of h_{FE} in the latter. This fact has been mentioned earlier.

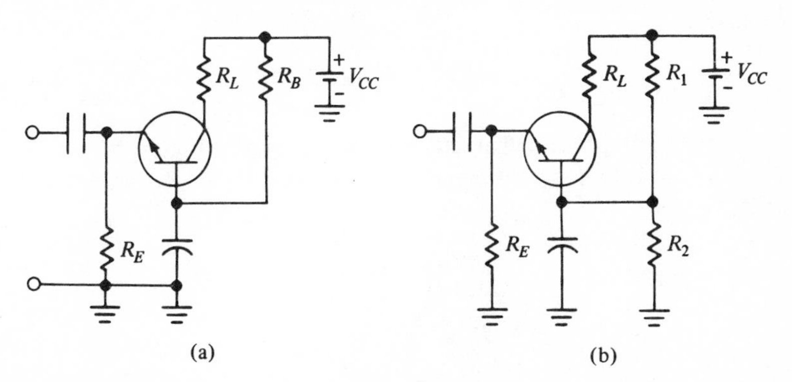

Figure 5–27 (a) CB simple biasing. (b) CC voltage-divider emitter-resistance biasing.

5-7 DC Operating Point

As stated previously, a transistor must have dc potentials before it can function as an amplifier. Secondly, these dc potentials must be such that the transistor does not conduct excessively or that the amplified signal is not distorted. When this operating point is chosen, circuit elements, such as batteries and resistors, are chosen to achieve the desired dc potentials. In all previous example problems the operating point had been chosen. It is now appropriate to see how this is done.

An output family of curves is used to illustrate this concept more dramatically. Since these are output curves it certainly would be advantageous if the load resistance could be superimposed on this curve. This can easily be done if we use the relationship,

$$V_{CC} = I_C(R_L + R_E) + V_{CE} \tag{5-72}$$

If we assume I_C to be zero in this equation, then

$$V_{CE} = V_{CC} \tag{5-73}$$

This is one terminal point for our load resistance. Since R_L is linear we need only one other point. We obtain it by assuming that $V_{CE} = 0$. Then Equation (5-72) becomes

$$V_{CC} = I_C(R_L + R_E) \tag{5-74}$$

or

$$I_C = \frac{V_{CC}}{R_L + R_E} \tag{5-75}$$

This is our other terminal point. If we draw a line between these two points we will have $(R_L + R_E)$ plotted on the family of curves. One point is $V_{CE} = 0$ and $I_C = V_{CC}/(R_L + R_E)$ and the other point is $I_C = 0$ and $V_{CE} = V_{CC}$.

EXAMPLE 19

A CE amplifier's characteristics are shown in Figure 5-28. If $R_L = 4.0\text{ k}\Omega$, $R_E = 1.0\text{ k}\Omega$, and $V_{CC} = 20$ volts, plot $R_L + R_E$ on a family of curves.

SOLUTION

$$I_C = \frac{V_{CC}}{R_E} = \frac{20}{5\text{ k}\Omega}$$
$$= 4\text{ mA}$$

Therefore, one point is at $V_{CE} = 0$ and $I_C = 4\text{ mA}$; the other point is at $I_C = 0$ and $V_{CE} = 20$ volts. On Figure 5-28, see line *A*.

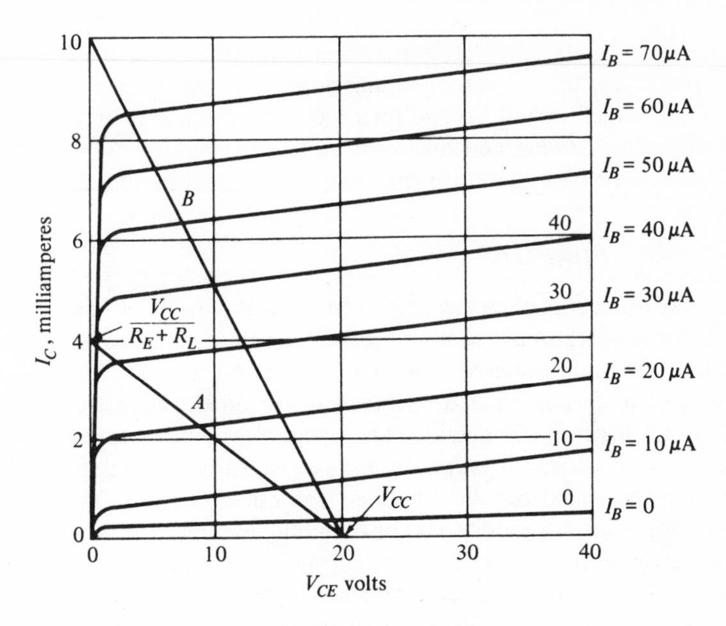

Figure 5–28 Load-line plots on CE characteristics.

EXAMPLE 20

Suppose a 20-volt supply is used and the circuit is such that $R_L = 2.0\ \text{k}\Omega$ and $R_E = 0$. Plot $R_L + R_E$ on the curve.

SOLUTION

$$I_C = \frac{V_{CC}}{R_L} = \frac{20}{2\ \text{k}\Omega}$$
$$= 10\ \text{mA}$$

See line *B*.

Analysis of line *A* tells us that V_{CE} can have any value from 0 to 20 volts depending on what value of I_B is chosen. Or, conversely, I_C can have any value from 0 to 4.0 mA, depending on what value I_B has. An I_B is usually chosen for class A amplifiers for the center of these ranges. In this case, $I_B = 17\ \mu\text{A}$. The intersection of the resistance line and the chosen I_B point is called the quiescent point because these are the current and voltage (I_C, V_{CE}) present when no signal is present. Now if no family of curves were available a quick mathematical manipulation would show that the maximum current for the chosen quantities (V_{CC}, R_E, and R_L) is $20/5\ \text{k}\Omega = 4.0\ \text{mA}$. Therefore, an appropriate collector current should be half this quantity, or 2.0 mA. If the h_{FE} is known, then I_B for proper operation would be $I_B = 2.0\ \text{mA}/h_{FE}$. Once I_B is known, the proper base-circuit resistances are calculated to obtain it.

One final word about line A. When a signal appears at the input, I_B changes above and below this chosen operating point. Correspondingly, I_C and V_{CE} change and hopefully not beyond their maxima or below their minima. That is why the signal swing and quiescent point are chosen so they stay within these limits. The best guarantee is to have a quiescent point near the center.

5-8 Summary

In this chapter we became familiar with the three basic transistor amplifier configurations: common base, common emitter, and common collector. All three required dc potentials before any ac signal could be amplified. The dc potentials determined where on the characteristic output curves the transistor will operate. The operating point essentially is chosen so that the transistor ratings (power and current) are not exceeded and at the same time the signal can be amplified without distortion. Therefore, this requires obtaining laboratory data or obtaining manufacturers' specifications before a given transistor can be used. Throughout the chapter we used simple circuit theory to select dc biasing resistors for each configuration.

The ac characteristics that discriminate one configuration from another are the ac input resistance, the voltage gain, and the current gain. Relatively speaking, the input resistances are low, medium, and high for CB, CE, and CC circuits, respectively. Correspondingly, the estimated resistances are r_j, βr_j, and $\beta(r_j + R_E)$. Voltage gains for CE and CB are approximately equal and the CC gain is less than one. For all three configurations, $A_v = R_L/(r_j + R_E)$. However, R_E is used in the equation only when no capacitor appears across it, as in the case of a CC arrangement.

We learned that a stage gain is more meaningful than a simple transistor amplifier gain. The stage gain has been defined as a comparison (voltage or current) of what entered transistor two (or useful load) to what entered transistor one. Usually these gains are lower than transistor amplifier gains because dc biasing resistors reduced the voltage or current that entered T_2.

Finally, we learned how to plot a dc load line and pick a midpoint for a quiescent operating point. The dc load line was plotted on a set of output characteristics by determining the two extreme operating conditions: V_{CC}, $I_C = 0$; and $V_{CE} = 0$, $I_C = V_{CC}/(R_E + R_L)$. The ideal quiescent point for class A amplifiers was found to be half of the maximum collector current, which is $V_{CC}/(R_E + R_L)$. This leads us to a conclusion that one can readily select an operating point by the use of the foregoing expression, thereby eliminating the use of the family of curves.

Questions and Problems

1. What minimum forward-bias voltages will cause conduction in a silicon diode? In a germanium diode?

2. A CE amplifier uses a germanium *n-p-n* transistor as in Figure 5-5(a). Voltmeter readings are as follows: $V_{CC} = 9.0$, $V_{BB} = 6.0$, $V_{RL} = 5.0$, $V_{RB} = 5.7$ volts. Determine (a) V_{CE}, (b) I_C if $R_L = 5.0\ \text{k}\Omega$, (c) I_B if $h_{FE} = 50$, and (d) R_B.

3. (a) Sketch a simple CE amplifier with two dc sources: R_B and a load resistance R_L. Use a *p-n-p* transistor. (b) For the amplifier in part (a), assume $V_{CE} = 3$ volts, $V_{CC} = 12$ volts, and $I_C = 4.5$ mA. Determine the value of R_L. (c) In part (a), assume that $V_{BB} = 4.5$ volts. What R_B is required if the germanium transistor is biased at 50 μA? (d) What is h_{FE} for this transistor?

4. In Problem 2, determine the approximate voltage gain expected.

5. In Problem 2, determine the input resistance R_i.

6. For Figure 5-29 a *p-n-p* silicon transistor is used. If a family of curves indicates that half the maximum collector current is suitable to bias it, determine (a) maximum collector current ($= V_{CC}/R_L$), (b) appropriate quiescent collector current ($V_{CC}/2R_L$), (c) R_B if $h_{FE} = 35$, (d) V_{CE}, and (e) A_v.

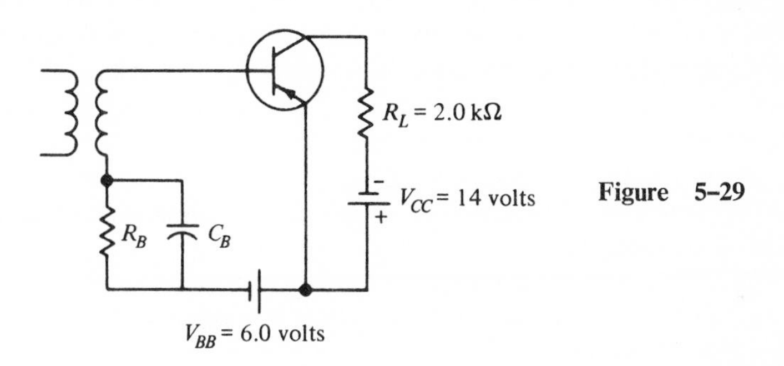

Figure 5–29

7. The equation for voltage gain ($A_v = R_L/r_j$) implies that increasing R_L increases gain. (a) Determine A_v for Problem 6 if $R_L = 4.0\ \text{k}\Omega$. Assume I_C is 3.5 mA. (b) Determine V_{CE} when $R_L = 4.0\ \text{k}\Omega$. (c) What conclusion can you reach on the value of the largest load resistance possible in a CE circuit?

8. Sketch the equivalent ac circuit for Figure 5-29.

9. A 10 μA peak signal is introduced to the amplifier in Problem 6. If $h_{fe} \simeq h_{FE}$, determine (a) $i_{C\,(\max)}$ and (b) $I_{C\,(\max)}$.

10. A transistor is chosen to operate at $I_C = 1.5$ mA. If the voltage gain is limited to 45, determine the appropriate load resistance.

11. What value of V_{CC} is appropriate in Problem 10 if V_{CE} must equal V_{R_L}?

12. Sketch a *p-n-p* common-base amplifier with appropriate batteries, bias resistance, and a transformer signal input.

13. Sketch the dc equivalent circuit for Problem 12.

14. In Figure 5-30, $V_{EE} = 6.0$ volts, $V_{CC} = 12.0$ volts, $R_E = 10\ \text{k}\Omega$, and $R_L = 6.0\ \text{k}\Omega$. If a germanium transistor is used, determine (a) I_E, (b) I_C, (c) V_{CB}, and (d) the maximum possible collector current. (*Hint:* assume $V_{CE} = 0$.)

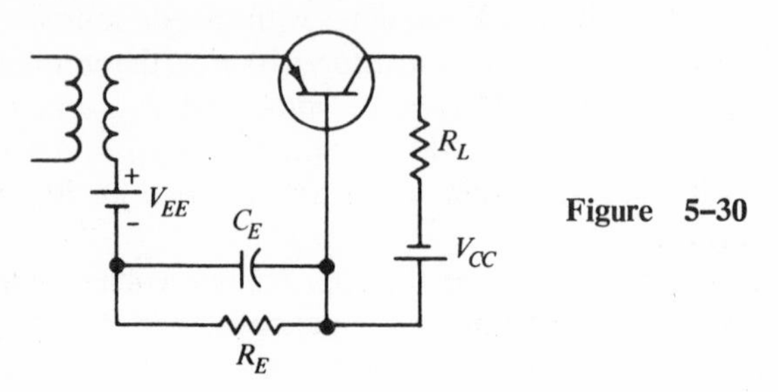

Figure 5-30

15. Determine the approximate ac input resistance for the conditions obtained in Problem 14.
16. The following dc voltages were measured in the circuit of Figure 5-30: $V_{CC} = 22$ volts, $V_{R_L} = 10$ volts, $V_{EE} = 12$ volts, $V_{R_E} = 11.3$ volts. If the collector current is 5.0 mA, determine (a) R_L, (b) I_E, and (c) R_E.
17. Assume that R_L in Problem 16 is increased to 4.0 kΩ. (a) Will the collector current change? (b) What is the dc voltage across R_L? (c) What is V_{CE}? (d) Is this an appropriate V_{CE}?
18. Determine the approximate voltage gain for Problem 14.
19. For Problem 16, determine (a) the approximate voltage gain and (b) the input resistance.
20. Sketch the ac equivalent circuit for Figure 5-30.
21. Sketch a CC amplifier, using a *p-n-p* transistor with the necessary load and bias resistors and batteries.
22. Using Figure 5-20(a), with $R_B = 20\ \text{k}\Omega$, $V_{BB} = 10$ volts, $R_E = 1.0\ \text{k}\Omega$, $V_{CC} = 12.0$ volts, and $h_{FE} = 50$, determine (a) I_E, (b) r_j, (c) R_i, (d) $I_E R_E$.
23. In the circuit of Figure 5-31, the following dc voltages were measured: $V_{CC} = 12$, $V_{BB} = 10$, $V_{R_E} = 1.0$ volts. If I_C equals 2 mA and h_{FE} equals 100, determine (a) R_E and (b) R_i.
24. For Problem 22, what is the voltage gain?

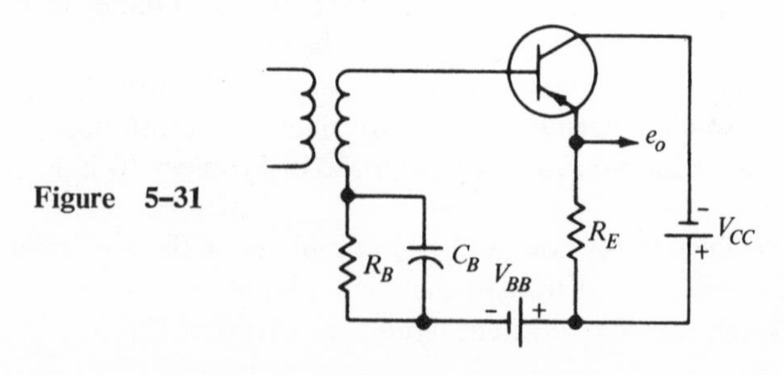

Figure 5-31

25. For Problem 23, what is the voltage gain?
26. If a microphone has an internal impedance of 50 kΩ, which configuration would you use to obtain maximum possible voltage transfer from the microphone? Assume that the connections are made directly (through a capacitor) to inputs of each configuration. Why? If you had a choice of a CB or CE arrangement, which would you choose? Why?
27. Using Figure 5-22(a), determine the stage gain if $r_{j1} = r_{j2} =$ 50 ohms, $R_{L1} = R_{L2} = 3.0$ kΩ, $\beta_1 = \beta_2 = 60$, and $R_{B2} = 50$ kΩ.
28. In Problem 27, R_{L1} is increased to 5.0 kΩ to increase the gain. What is the new gain?
29. In Figure 5-32, assume that the 5.0 kΩ is the lumped resistance of all resistors that appear beyond C_C. If $I_E = 2.1$ mA, $V_{CC} = 9.0$ volts, and $h_{FE} = 75$, determine (a) r_j, (b) V_{CE}, (c) I_B, (d) A_v (stage gain), and (e) R_i.

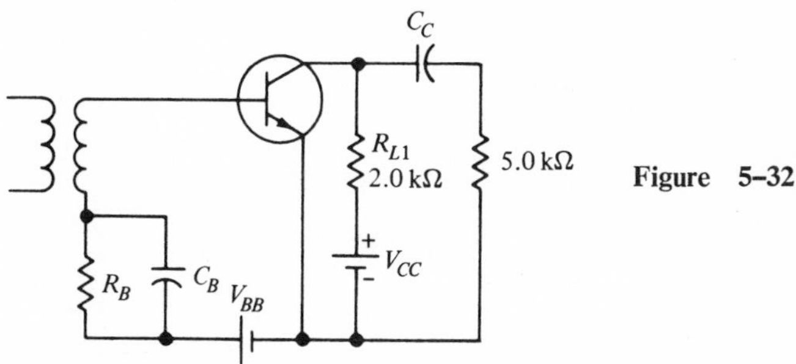

Figure 5–32

30. In Figure 5-32, determine the current gain. Assume that the 5.0 kΩ is R'_o.
31. Determine A_i for Problem 27 if R'_o is $r_{j2}B_2$. Compare this to B_1.
32. In Figure 5-33, $V_{CC} = 12.0$ volts and the desired I_B is 120 μA. What value of R_B is required?
33. If $I_C = 2.0$ mA, $h_{FE} = 50$, and $V_{CC} = 12.0$ volts in Figure 5-33, determine the value of R_B required.
34. In the Problem 33, determine (a) R_i and (b) the parallel equivalent of R_i and R_B.
35. In Figure 5-25, $I_C = 1.0$ mA, $I_B = 20$ μA, $R_E = 500$ ohms, $V_{CC} =$ 12 volts. If we assume that the silicon transistor permits a bleeder current (I_2) of $5I_B$, determine R_1 and R_2.

Figure 5–33

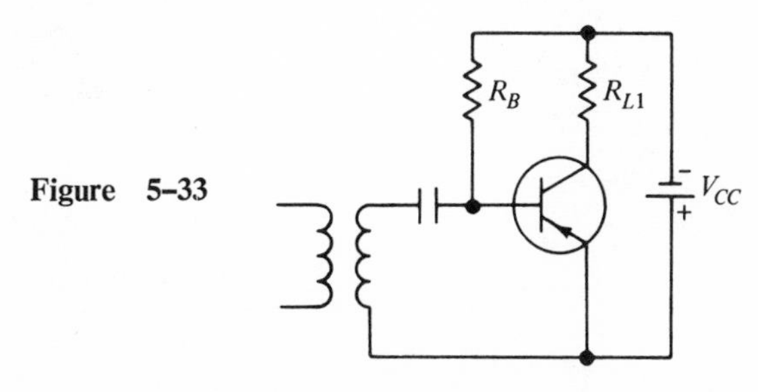

36. In Problem 35, solve for R_i when (a) R_E is bypassed by a large capacitor and (b) R_E is not bypassed. (*Hint:* use $\beta(r_j + R_E$.)
37. On Figure 5-28 plot the following load resistances if $V_{CC} = 40$ volts: (a) 20 kΩ, (b) 8 kΩ.
38. What are the maximum currents that the collector may see in Problem 37?
39. Select the appropriate I_B for the two loads in Problem 37.
40. Assume that a load resistance of 3.0 kΩ is connected to a battery of 6.0 volts in a CE amplifier. (a) What is the maximum possible current for the collector? (b) What is appropriate collector current? (c) If $h_{FE} = 100$, what should be the value of I_B for part (b)?

6

The Field-Effect Transistor

6-1 The Junction Field-Effect Transistor (JFET)

The field of semiconductors has expanded considerably because many devices besides the ordinary transistor have appeared on the market. One of these is the field-effect transistor, commonly abbreviated FET. Its operation is not difficult to comprehend, particularly if the study of transistors is behind us.

If a block of semiconductor material is doped and terminals are placed on opposite ends, we can expect conduction to occur. The higher the impressed voltage, the greater will be the current flow. In this simple arrangement, shown in Figure 6-1, the semiconducting material is of the *n* type and the terminals are called source and drain. Just as we did in transistors, we have established current flow. If we want to make this electronic device useful, the next step is to establish a means of control over the moving particles. If we add two *p* regions on the side of the *n*-type material, we have the makings of an FET. Figure 6-1(b) shows these two *p* regions tied together and labeled *G*, representing the gate. Apparently we can expect, by attaching some significance to the word *gate*, that current flow between the source and the drain will be controlled by the gate. Indeed, this is precisely what happens. Figure 6-1(c) shows a battery V_{GG} connected between the gate and source, with the gate negative. Notice that only one gate is connected and this is only to simplify the diagram. Functionally, both gates are tied together. If the gate, which is *p* material, is negative with respect to the source, which is *n* material, we have a case of reverse bias. Reverse bias implies low current flow. However, not only is the current low for the gate to source diode but

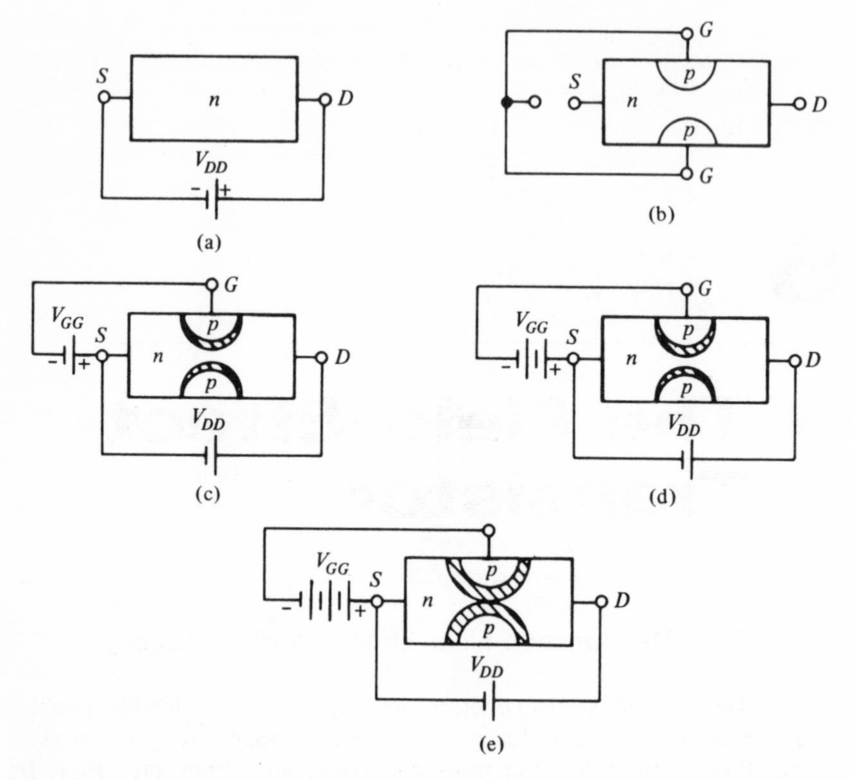

Figure 6–1 Schematic diagram showing depletion as reverse bias is increased.

also the current is reduced between the source and the drain. If the reverse bias is increased we increase the depletion region, which then decreases majority carriers that contribute to the source-to-drain current. Figure 6-1(d) shows the depletion region penetration for a larger reverse bias. Finally, Figure 6-1(e) shows the depletion region completely choking off the channel where the source-to-drain current flows. The particular gate voltage that cuts off the drain current is called pinch-off voltage. It is one of the discriminating features of FETs.

After studying the previous chapters one can anticipate that a signal entering the FET will affect the reverse bias rather minutely and correspondingly have a large effect on the output drain current.

It must be pointed out that not all FETs are *n*-channel. When the basic semiconductor is *p* material, we have a *p*-channel FET. Therefore, the gates must be *n*-doped regions, with the positive side of the battery going to the gate to maintain reverse bias between gate and source. The drain-to-source battery need not be reversed.

Summarizing, we can say that the FET, like the ordinary transistor, is a three-element device. If we must make analogies, the gate corresponds to the base, the source to the emitter, and the drain to the collector. The ordinary transistor requires one reverse-biased junction and another biased in the forward direction. On the other hand, the FET has only one junction, between the gate and source, and it is reverse-biased. This is why FETs are called unipolar devices and ordinary transistors bipolar. Finally, since the FET is a three-element device, we can expect that three different configurations can be obtained, with each producing its own characteristics.

6-2 The Ampere–Volt Characteristics of a JFET

Once again we go through the procedure that was established when transistors were discussed; that is, after some physical explanation of a device we proceeded through its static characteristics, then through dc, and finally ac circuit considerations. In the process we established some mathematical relationships that further enhanced our comprehension of the electronic device. Hopefully this pattern will also permit some analogies that will make understanding a little easier.

If we took Figure 6-1(c) and ran a test to see how the drain current was affected as we change the drain-to-source voltage for different levels of gate–source voltage, Figure 6-2 would result. Notice that if the gate–source voltage (V_{GS}) is zero the drain currents are relatively high as we increase drain–source voltage. If we introduce a negative voltage (gate negative with respect to source), the drain currents are at lower values as we proceed through the range of drain–source voltage. This shows graphically what was said in Section 6-1; that is, the depletion region penetrates deeper into the channel

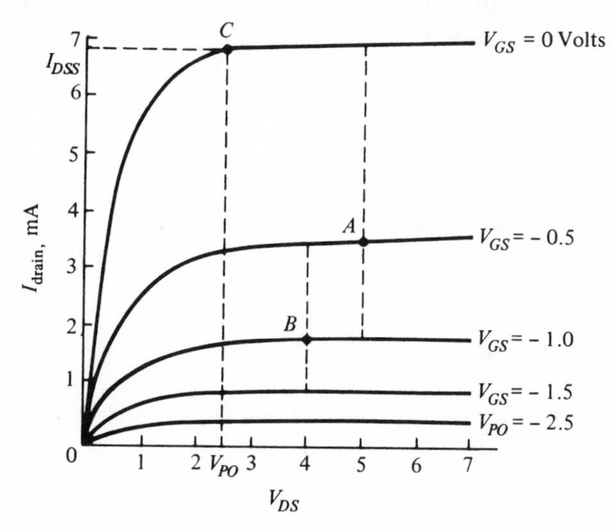

Figure 6–2 Typical drain family of curves.

and limits the carriers that are available for current flow in the drain circuit. If we further increase the gate to drain negative bias, the drain current is further reduced.

Let us step back and make an inspection of this family of curves. Notice that the vertical axis is drain current I_D and the horizontal axis is drain-to-source voltage, V_{DS}. If we were looking at a CE set of curves the vertical axis would be I_C and the horizontal axis would be V_{CE}. Therefore, can we not conclude that analogously we are looking at a common-source (CS) set of curves? Indeed so, because previous transistor experience has told us that these are output quantities and the second subscript of the voltage quantity indicated the common terminal. Continuing with this analogy, we expect that in the field of the curves an input quantity should appear. In the case of a CE configuration it was base current I_B. In the case of FETs we digress because the input quantity is not current but a voltage, V_{GS}. Despite this difference, the characteristic curves describe how output quantities are affected as an input quantity is changed.

A further look at these curves tells us that they are similar to CE curves except that they do not rise (from zero) as quickly as ordinary transistor curves do. This means that the "useful" voltage range for amplifier operation is limited. Putting it in another way, a quiescent or operating point will be selected in the flat portion of the curves and the curved portion will be avoided. Finally, we notice that CE and CS differ in the direction of ascending input quantities. For CE, the $I_B = 0$ curve is near the horizontal axis, whereas for the CS configuration the $V_{GS} = 0$ curve is at the top.

Just as we did in transistors we can also extract useful quantities off the FET family of curves. The most important quantity is transconductance (g_m). The student will recall that conductance implies a current/voltage ratio and in addition "trans" means across. Therefore, "across" the FET with a I/V ratio automatically means I_D/V_{GS}. To extract this quantity from the curve, V_{DS} must be held constant. Mathematically,

$$g_m = \left.\frac{\Delta I_D}{\Delta V_{GS}}\right|_{I_D=\text{constant}} \tag{6-1}$$

This type of operation was done on ordinary transistor curves when we extracted beta or alpha. The following example is used to demonstrate Equation (6-1) and to show the variability in the quantity.

EXAMPLE 1

Determine g_m near (a) point A and (b) point B on Figure 6-2.

SOLUTION

(a)

$$g_m = \left.\frac{\Delta I_P}{\Delta V_{GS}}\right|_{I_D=\text{constant}}$$

Following the dotted lines, we obtain

$$g_m = \frac{(7.0 - 1.8)(10^{-3})}{1.0 - 0}$$

$$= 5.2 \times 10^{-3} \text{ mho}$$

(b)

$$g_m = \frac{(3.5 - 0.9)(10^{-3})}{1.5 - 0.5}$$

$$= 2.6 \times 10^{-3} \text{ mho}$$

From this example and a look at the curves we can conclude that g_m will be small near the bottom and relatively larger near $V_{GS} = 0$. Since g_m takes on this range of values, manufacturers usually quote typical, minimum, and maximum values of g_{m0}, which is the transconductance at $V_{GS} = 0$.

Another quantity that can be obtained from the curves is r_{DS}, which is the drain resistance. This is similar to collector resistance. It too is a large quantity relative to other resistors that are used in a circuit and for a quick approximation it is usually omitted. To verify the magnitude of r_{DS}, a quick look at Figure 6-2 and the equation defining r_{DS} will be sufficient. Mathematically,

$$r_{DS} = \left.\frac{\Delta V_{DS}}{\Delta I_D}\right|_{V_{GS}=\text{constant}} \tag{6-2}$$

Figure 6-2 will yield a large ΔV_{DS} and a small ΔI_D along any V_{GS} in the flat region. The net result is that r_{DS} will be large.

A final look at Figure 6-2 tells us that the pinch-off voltage, V_{PO}, is -2.5 volts for this FET. Usually this voltage is not easily obtained by increasing the bias until drain current is essentially zero. This is so because the drain current does not become zero abruptly but slowly tapers to very low values. A much easier way to obtain V_{PO} is to determine the drain-to-source voltage at which drain current becomes essentially constant when $V_{GS} = 0$; see point C on Figure 6-2. The corresponding drain current is I_{DSS}, which means the drain-to-source current (I_{DS}), with the third element short-circuited (I_{DSS}). The third element is the gate, and short-circuiting the gate to the source means that $V_{GS} = 0$. This may be a bit confusing, especially if we are obtaining a drain–source voltage and calling it a pinch-off voltage, which is a gate–source quantity. The matter can be explained. There is a relationship between the three voltages V_{PO}, V_{DS}, and V_{GS}, where each curve begins to level out.

$$V_{PO} = V_{DS} + |V_{GS}| \tag{6-3}$$

This is where $|V_{GS}|$ is the absolute value (ignore the sign) of V_{GS}. This means that if V_{DS} increases then V_{GS} must decrease and vice versa to maintain the pinch-off voltage for a given FET. Therefore, if we make $V_{GS} = 0$ in this equation, then $V_{PO} = V_{DS}$. This is shown as point C in Figure 6-2.

At this point we must backtrack and summarize. The FET conducts when majority carriers are subject to a potential between two elements, source and drain. Control over these carriers is achieved when the gate, which is opposite in doping to the channel, has a reverse bias applied between it and the source. The greater the reverse bias, the smaller will be the channel current. When raised sufficiently high the drain current is essentially zero. This particular gate-to-source voltage is called the pinch-off voltage. The drain current that flows when the gate voltage is zero, and pinch-off magnitude is across the drain and source, is called I_{DSS}. The final characteristic of importance is the transconductance, which essentially tells us how much the drain current will change for a change in gate–source voltage.

To conclude this section we reveal a dilemma. Most FET parameters are quoted as having a range of values. We already mentioned g_{m0} as one; others are V_{PO} and I_{DSS}. The dilemma is this: if these quantities are needed in mathematical relationships, what value is appropriate—typical, minimum, or maximum? Ordinary transistors have some of this variability. In either case, probably the simplest way out is to measure the quantities of interest for each transistor or FET that is to be used. The next section will show how this is done.

6-3 Measurement of Transconductance and I_{DSS}

In FETs, the following simple equation yields the voltage gain:

$$A_v = g_m R_L \tag{6-4}$$

where g_m is the transconductance near the region of operation. We shall see that g_m is related to g_{m0}. To be precise, we shall measure g_{m0} instead of relying on the manufacturer's range of values.

To begin, we introduce the symbol for n-channel FET in Figure 6-3(a). Notice that the gate carries a transistor symbol for p material. Therefore, if the gate is of the p-type, the bulk material or channel must be n. Conversely, Figure 6-3(b) shows a p-channel FET because of the n-type gate. Finally, there is no distinction between the drain and source; hence they can be used interchangeably.

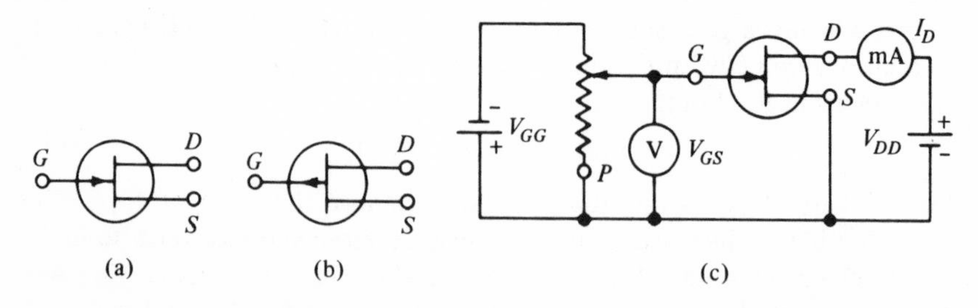

Figure 6–3 (a) Symbol for n-channel FET. (b) Symbol for p-channel FET. (c) Circuit used for determining I_{DSS} and g_{m0}.

Figure 6-3(c) shows a simple test circuit that can be used to extract the necessary parameters. Notice that the potentiometer in the gate circuit permits us to bias the gate to any negative value as we move the arm. If the arm is at position P, then the gate is short-circuited to the source. If V_{DD} is now raised to any value greater than the anticipated pinch-off voltage, or to where the current does not increase appreciably, the milliammeter will indicate I_{DSS}. This indeed is a condition where the drain-to-source current is monitored while the gate is short-circuited ($V_{GS} = 0$).

If we now place some bias on the gate (moving the arm up), the drain current will obviously decrease. Hence, we have a condition of a change in I_D while the drain–source voltage remained constant. Thus

$$g_{m0} = \left.\frac{\Delta I_D}{\Delta V_{GS}}\right|_{V_D=\text{constant}} \tag{6-5}$$

These changes must be small but perceptible. The transconductance obtained in this manner is the transconductance near the region of $V_{GS} = 0$. This is why the symbol is g_{m0}.

It can be mentioned that the circuit of Figure 6-3(c) can also be used to plot the family of curves. V_{GS} is first selected, and I_D is plotted for corresponding V_{DS}'s. The process is repeated for another V_{GS}, and so forth.

The initial intention for the circuit was to obtain I_{DSS} and g_{m0} and with these we hopefully expect to use them in mathematical relationships, just as we used beta and r_j in our study of ordinary transistors to determine gain and input resistance. Suffice it to say that FET characteristics—even those within the same family—vary considerably. The quickest way out is to measure these characteristics in a simple circuit and then proceed using the measured quantities with some degree of confidence. We can conclude with the following example:

EXAMPLE 2

A test circuit as shown in Figure 6-3(c) produced the test data shown in Table 6-1.

Table 6-1

Test	V_{GS}	V_{DD}	I_D
1	0	5	2.0 mA
2	−0.1	5	1.85 mA

Determine (a) I_{DSS} and (b) g_{m0}.

SOLUTION

(a) $$I_{DSS} = 2.0 \text{ mA}$$

(b) $$g_{m0} = \left.\frac{\Delta I_D}{\Delta V_{GS}}\right|_{V_D=\text{constant}}$$

$$= \frac{(2.0 - 1.85)(10^{-3})}{0.1 - 0} = \frac{0.15 \times 10^{-3}}{0.1}$$

$$= 1.5 \times 10^{-3} \text{ mho}$$

6-4 The Common-Source Amplifier

The common-source amplifier is similar to the common-emitter amplifier. One big difference between the two is that the former has a much higher input resistance. One must recall that in a CE amplifier we require forward bias between the input terminals (base to emitter). This means a relatively small resistance. The CS amplifier, on the other hand, requires reverse bias between input terminals (gate to source). This, in turn, means relatively high resistance because this leakage current is in the microampere range—implying an input resistance in megohms. (One can envision the usefulness if the microphone problem of Chapter 5 is recalled.)

Figure 6-4 shows a simple CS amplifier using two dc sources. As in transistors, we must satisfy the dc conditions first. In the input side, however, we

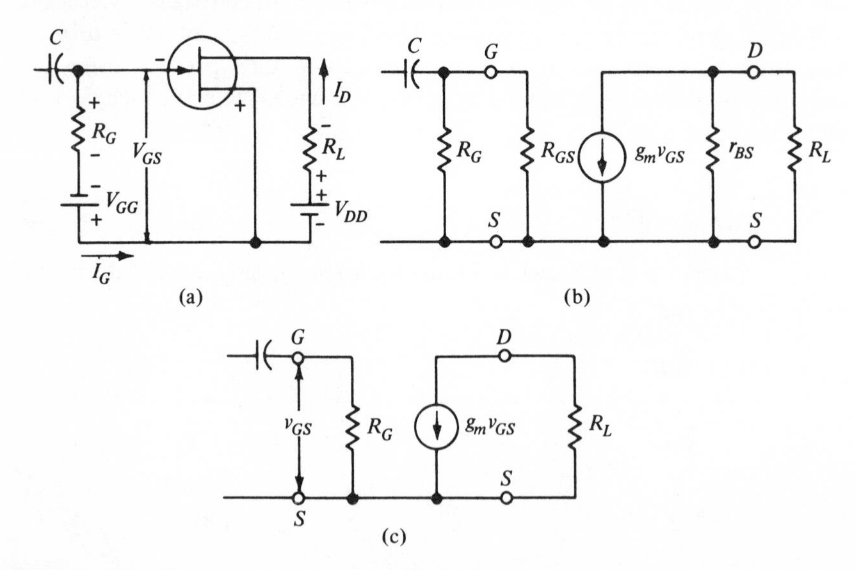

Figure 6-4 Common-source (a) amplifier, (b) equivalent circuit, and (c) simplified equivalent circuit.

must design for an operation voltage V_{GS}, and not a current as we did with ordinary transistors. The dc equation for the gate circuit is

$$-V_{GG} + V_{GS} + I_G R_G = 0 \tag{6-6}$$

or

$$V_{GS} = V_{GG} - I_G R_G \tag{6-7}$$

It is evident that if $I_G R_G$ is small enough we can expect V_{GS} to be approximately equal to the battery voltage V_{GG}. Keep in mind V_{GS} is one of the voltages that determines the operating point and it is selected by the user or designer. Since I_G is the leakage current and it is in the microampere range we might expect $I_G R_G$ to be small. However, since I_G is reverse current we may expect a relatively large value if the temperature is raised. Also, R_G is in the megohm range primarily because of ac conditions that usually require high input resistance. (Once again, remember the microphone example.) In any case, one must remember that the selected battery for this circuit has to be greater than the selected operating V_{GS}.

In the output side, the dc equation is similar to the CE equation:

$$-V_{DD} + I_D R_L + V_{DS} = 0 \tag{6-8}$$

or

$$V_{DS} = V_{DD} - I_D R_L \tag{6-9}$$

This means that selection of dc voltage, load resistance, and drain current must be such that the FET drain–source voltage is in the flat region of operation. Depending on the unknown quantities, a rough rule of thumb is to select an I_D equal to half of the maximum I_D $(= V_{DD}/R_L)$ expected for the chosen load resistance. A few examples are appropriate.

EXAMPLE 3

An FET as shown in Figure 6-4 was chosen to operate with a drain current of 4 mA and a $V_{GS} = 0.5$. (a) If $R_L = 2.0\ \text{k}\Omega$, determine the supply voltage required if V_{DS} is supposed to be approximately equal to the dc drop across the load resistance. (b) If $R_G = 1.0$ M ohms, what value of V_{GG} is required if a leakage of 1.0 μA is expected?

SOLUTION

(a)

$$\begin{aligned} I_D R_L &= 4.0(10^{-3})(2)(10^{3}) \\ &= 8 \text{ volts} \end{aligned}$$

$$\begin{aligned} V_{DD} &= I_D R_L + V_D \\ &= 8 + 8 \\ &= 16 \text{ volts} \end{aligned}$$

(b)

$$I_G R_G = (1.0)(10^{-6})(1)(10^6)$$
$$= 1.0 \text{ volt}$$

$$V_{GG} = V_{GS} + I_G R_G$$
$$= 0.5 + 1.0$$
$$= 1.5 \text{ volts}$$

EXAMPLE 4

An FET, as in Figure 6-4, has a load resistance of 10 kΩ and a V_{DD} of 20 volts. Determine an approximate operating drain current.

SOLUTION

$$I_{D\,(\max)} = \frac{V_{DD}}{R_L}$$
$$= \frac{20}{10 \text{ k}\Omega}$$
$$= 2 \text{ mA}$$

$$I_{DQ} \simeq \tfrac{1}{2} I_{D\,(\max)}$$
$$= \tfrac{1}{2}(2 \text{ mA})$$
$$= 1.0 \text{ mA}$$

Figure 6-4(b) shows the ac equivalent circuit of 6-4(a). R_{GS} is the reverse bias resistance offered by the gate–source diode. The output side is simulated by a current generator, $g_m V_{GS}$, because the FET characteristically behaves like a constant-current source. The student may verify by substitution to prove that $g_m V$ is a current. Since R_{GS} is much greater than R_G and similarly r_{DS} is much greater than R_L, we can further simplify the equivalent circuit. This is shown in Figure 6-4(c). This circuit permits us to extract a voltage-gain equation. By definition,

$$A_v = \frac{e_o}{e_i} \tag{6-10}$$

If e_o is the voltage that appears across R_L, and the current supplied by the generator passes through R_L, then

$$e_o = g_m v_{GS} R_L \tag{6-11}$$

Also, if

$$e_i = v_{GS} \tag{6-12}$$

then

$$A_v = \frac{g_m v_{GS} R_L}{v_{GS}} = g_m R_L \tag{6-13}$$

Equation (6-13) is quite simple. The only drawback is determining g_m. It can be obtained from a family of curves if it is supplied or obtained by test in a laboratory. Recall that this value of g_m must be extracted at the operating point because g_m depends on the region of operation. We can also solve for g_m mathematically once I_{DSS} and g_{m0} are obtained. The following equations are appropriate if this latter route is taken. If I_{DSS} and g_{m0} are known, then

$$V_{PO} = \frac{2I_{DSS}}{g_{m0}} \tag{6-14}$$

If we know V_{PO}, then g_m at any operating point is

$$g_m = g_{m0}\left(1 - \frac{|V_{GS}|}{V_{PO}}\right) \tag{6-15}$$

EXAMPLE 5

A test on an FET produced an I_{DSS} of 2.0 mA and a g_{m0} of 2000 μmho. If load resistance of 10 kΩ is chosen determine the gain when V_{GS} is (a) -1.0, (b) -0.5, (c) 0 volts.

SOLUTION

$$\begin{aligned} V_{PO} &= \frac{2I_{DSS}}{g_{m0}} \\ &= \frac{2(2.0)(10^{-3})}{2(10^{-3})} \\ &= 2.0 \text{ volts} \end{aligned}$$

(a)

$$\begin{aligned} g_m &= g_{m0}\left(1 - \frac{V_{GS}}{V_{PO}}\right) \\ &= 2000(10^{-6})(1 - \tfrac{1}{2}) \\ &= 1000\ \mu\text{mho} \\ A_v &= g_m R_L \\ &= 1000(10^{-6})(10)(10^3) \\ &= 10 \end{aligned}$$

(b)

$$\begin{aligned} g_m &= 2000(10^{-6})\left(1 - \frac{0.5}{2}\right) \\ &= 2000(10^{-6})(.75) \\ &= 1500\ \mu\text{mho} \\ A_v &= g_m R_L \\ &= 1500(10^{-6})(10)(10^3) \\ &= 15 \end{aligned}$$

(c)

$$g_m = 2000(10^{-6})\left(1 - \frac{0}{2}\right)$$
$$= 2000(10^{-6})$$
$$A_v = g_m R_L$$
$$= 2000(10^{-6})(10)(10^3)$$
$$= 20$$

Example 5 proved what was mentioned before—that the g_m is largest when $V_{GS} = 0$ and consequently a larger voltage gain results. Thus, if we choose to bias at $V_{GS} = 0$, the signal that is superimposed on this dc bias will drive the gate positive on its positive half-cycles. This situation, of course, is not desirable because we would be forward-biasing the gate–source diode, which in turn means decreased resistance. Therefore, we would be nullifying the advantage of high input resistance for FETs.

To characterize the FET further let us take a closer look at the input. An FET with its reverse bias at the input terminals may have a resistance of 10^9 ohms. At elevated temperatures this resistance may drop to the order of 10^7 ohms. Placing this resistance across typical R_G's has very little effect. Therefore, the resistance that a previous stage sees looking into the transistor stage is essentially R_G.

On the output side, it is necessary to evaluate the ac load seen by the current generator, which is comparable to the load resistance in ordinary transistors. The gain equation for FET may be appropriately rewritten as

$$A_v = g_m Z_L \tag{6-16}$$

where Z_L represents all the resistances that are connected across the drain and source. This load includes the FET's load resistance, all biasing, and the input resistance of a subsequent stage.

EXAMPLE 6

A CS amplifier has a 20-kΩ load resistance. If the signal is fed through a low-reactance capacitor and then to an ordinary transistor whose effective input resistance is 20 kΩ, determine the voltage gain of the FET. The operating conductance is 2000 μmho.

SOLUTION

$$Z_L = \frac{1}{(1/20\ \text{k}\Omega) + (1/20\ \text{k}\Omega)}$$
$$= 10\ \text{k}\Omega$$
$$A_v = g_m Z_L$$
$$= 2000(10^{-6})(10)(10^3)$$
$$= 20$$

6-5 The Common-Drain Amplifier (Source Follower)

A common-drain amplifier is simply a common source amplifier with its load resistance moved into the source lead. This same technique was used in transistors when we converted a CE amplifier to a CC amplifier. We also recall that in the CC amplifier we realized a higher input impedance and a low voltage gain (less than one). One can readily anticipate similar results in a common-drain amplifier except that the resistance is "moved up" into the megohm range.

Figure 6-5(a) shows a simple common drain or, as it is commonly called, a source follower. For dc requirements we bias in the middle of the flat portion of the family of curves as we did for a CS amplifier. Similarly, on the output side, the dc supply is dropped across the drain to source and across the unbypassed source resistance. Mathematically,

$$V_{DD} = V_{DS} + V_{RS} \tag{6-17}$$

or

$$V_{DD} = V_{DS} + I_D R_S \tag{6-18}$$

Because the source resistor is unbypassed, it becomes part of the input circuit, as shown by the equivalent circuit given in Figure 6-5(b). We must recall that in a common-drain amplifier, the input is between gate and drain and the output is between source and drain. This means that the gain is a comparison between these two quantities. Hence,

$$A_v = \frac{e_o}{e_i} = \frac{e_{SD}}{e_{GD}} \tag{6-19}$$

Figure 6-5 reveals that

$$e_{SD} = i_D R_S \tag{6-20}$$

$$= g_m v_{GS} R_S \tag{6-21}$$

If we make an ac loop equation at the input, the gate-to-drain voltage is equal to the generator voltage:

$$e_G = v_{GS} + e_{SD} \tag{6-22}$$

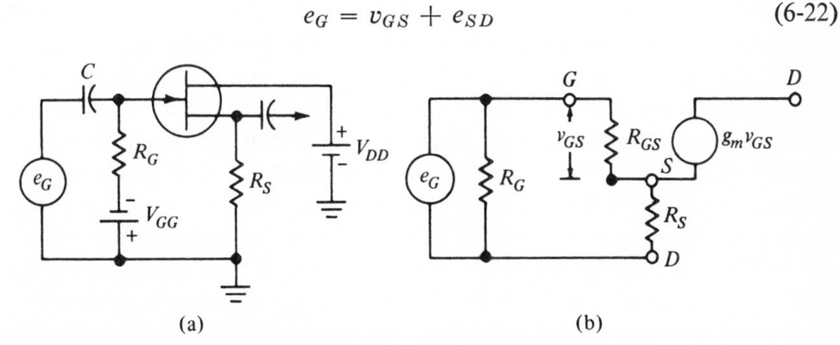

Figure 6–5 (a) Common-drain amplifier. (b) Equivalent circuit.

or

$$e_{GD} = v_{GS} + g_m v_{GS} R_S \tag{6-23}$$

It is this equation that tells us that R_S becomes part of the input circuit. To obtain a voltage-gain equation, we substitute Equations (6-23) and (6-21) into (6-19):

$$A_v = \frac{g_m v_{GS} R_S}{v_{GS} + g_m v_{GS} R_S} \tag{6-24}$$

$$= \frac{g_m R_S}{1 + g_m R_S} \tag{6-25}$$

A close look at this equation tells us that the gain will always be less than unity. Also, we need to know g_m at the operating point to extract a reasonable value for voltage gain.

Just as we realized in a CC amplifier, we expect the unbypassed source resistor to reflect a high resistance to the input. Indeed it does, but placing this high resistance across a much lower R_G does not increase the net input resistance appreciably. The input resistance for this kind of circuit remains essentially at the value of R_G. It, of course, is chosen for dc considerations. In any case, in comparison to a CC circuit, the FET source follower has a considerably higher input resistance. There are circuits that can cancel the effect of R_G at the input to yield megohms of input resistance. These are beyond the scope of this text. However, as a point of information this technique is called bootstrapping.

Notice the similarity of Equation (6-25) to the gain equation for a common source. Both have the product of g_m and load resistance. ($R_S = R_L$ in a common-drain circuit.) The chief difference is that for a common drain this product must be divided by a factor, $1 + g_m R_S$. This factor is a number greater than one and, when divided into $g_m R_S$, yields a gain less than one. In fact, this very same factor is used to determine what effective output resistance the source follower has. Mathematically,

$$R_o = \frac{R_S}{1 + g_m R_S} \tag{6-26}$$

The output resistance is mentioned because very frequently an FET is used to feed an ordinary transistor circuit. This being the case, the device that feeds an ordinary transistor must have relatively low impedance because these transistors themselves have low values of input resistance. Hence, the source follower offers relatively low generator impedance for a driven transistor. At the same time the FET source follower takes advantage of its high input impedance. A classical example of this situation is the solid-state "vacuum-tube voltmeter." By nature the VTVM must have a high input impedance. With the appearance of FETs on the market, the solid-state devices have challenged the vacuum tubes because the range of input impedances have now reached the tube range. Prior to this, ordinary transistors could not do the job. Once the input impedance demand was met with FETs, ordinary transistors were used in the following circuit. This is the reason matching FET output impedance with transistor input impedance is mentioned.

EXAMPLE 7

A source-follower circuit has an operating drain current of 5.0 mA, a g_m of 1500 μmho, and an R_S of 500 ohms. Determine the values of (a) A_v, (b) R_o, and (c) V_{DD} required if $V_{RS} = V_{DS}$.

SOLUTION

(a)

$$A_v = \frac{g_m R_S}{1 + g_m R_S}$$

$$= \frac{2000(10^{-6})(500)}{1 + 2000(10^{-6})(500)}$$

$$= 0.5$$

(b)

$$R_o = \frac{R_S}{1 + g_m R_S}$$

$$= \frac{500}{1 + 2000(10^{-6})(500)}$$

$$= 250 \text{ ohms}$$

(c)

$$V_{RS} = I_D R_S$$

$$= 5.0(10^{-3})(500)$$

$$= 2.5 \text{ volts}$$

$$V_{DD} = 2.5 + 2.5$$

$$= 5.0 \text{ volts}$$

6-6 Biasing the FET

Once again it would be advantageous to use only one dc source instead of two, as in the case of ordinary transistors. Figure 6-6 shows biasing using the self-bias approach in a CS amplifier. The voltage polarity across R_S is shown. This voltage tends to make the gate negative with respect to the source—precisely what is desired, because a reverse-bias demand requires that the *p* gate be negative. It is necessary to keep R_G in the circuit because it serves as a dc medium through which this bias voltage can be applied. Once again, we cannot ignore the dc voltage across R_G if leakage current flows through it. Hence,

$$V_{GS} = I_D R_S - I_G R_G \tag{6-26}$$

Here we see that $I_G R_G$ tends to reduce the negative bias. If elevated temperatures are anticipated, then temperature-dependent I_G will increase and

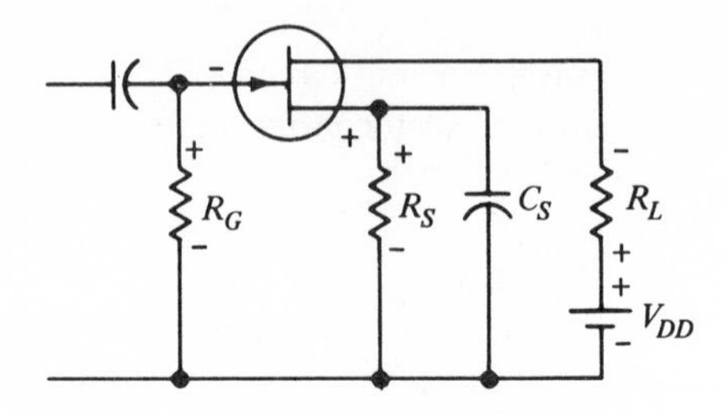

Figure 6–6 Common-source amplifier with self-biasing.

change the designed operating point. It is interesting to follow up on the net result if I_G increases. An increase in I_G causes $I_G R_G$ to increase. From Equation (6-26), V_{GS} will be reduced. Reducing V_{GS} means that the negative bias is less and therefore more current will flow in the drain circuit. More drain current means a larger $I_D R_S$ in Equation (6-26). This in effect cancels the increase of $I_G R_G$. Hence, where thermal runaway took place in transistors, FETs tend to protect themselves.

Capacitor C_S bypasses R_S, so R_S has no ac effect. If R_S were left unbypassed in a CS amplifier, the gain would be reduced, as we might expect. The reduction in gain is by a factor of $1/(1 + g_m R_S)$, or

$$A_v = \frac{1}{1 + g_m R_S} g_m R_L \tag{6-27}$$

Notice the similarity between this equation and the source-follower equation, wherein R_S had to be unbypassed and R_L was equal to zero. Despite the fact that the gain is reduced with an unbypassed source resistor, amplifiers purposely are designed with such conditions because other benefits are achieved. One is less distortion in the output.

As a final note, junction FETs have characteristics similar to, and behave very much like, vacuum tubes. The inputs of both are voltage-controlled—as opposed to ordinary transistors, which are current-controlled. The equations for gain are identical. Therefore, the study of FETs has laid the groundwork for tube study, which will appear in a later chapter.

EXAMPLE 8

An amplifier, as shown in Figure 6-6, requires a V_{GS} of 2.0 volts if a drain current of 1.0 mA is desired. (a) Determine the value of R_S if $I_G R_G$ is ignored. (b) If $R_L = 10\ \text{k}\Omega$ and $V_{DS} = 4.0$ volts, determine V_{DD}.

(a)

$$R_S = \frac{V_{GS}}{I_D} = \frac{2.0}{1.0(10^{-3})} = 2\ \text{k}\Omega$$

(b)

$$
\begin{aligned}
I_D R_L &= 1(10^{-3})(10)(10^3) \\
&= 10 \text{ volts} \\
V_{DD} &= I_D R_L + V_{DS} + I_D R_S \\
&= 10 + 4 + 2 \\
&= 16 \text{ volts}
\end{aligned}
$$

6-7 The Insulated-Gate FET

Section 6-1 gave a pictorial view of a junction FET with two gates on opposite sides of the bulk material. Instead of diffusing gates on opposite ends of the bulk material, manufacturers use a single-end approach as shown in Figure 6-7(a). Gate 2 is simply the *p*-type bulk material called the substrate. The narrow region between this substrate and gate 1 is the *n*-type channel. Figure 6-7(b) shows an insulating layer of silicon dioxide (SiO_2) beneath gate 1. This type of FET is called an insulated-gate FET or IGFET. To be more specific, it is a metal oxide semiconductor and therefore more commonly abbreviated as MOSFET.

The question arises, what kind of characteristics result from this arrangement? First, one can conclude that the reverse current between gate and source will be smaller and therefore the input resistance is greater than for the junction FET. The input resistance is in the area of 10^{15} ohms. Second, we can expect to use not only negative bias between gate and source, but also positive bias because the insulating layers prohibit forward-bias current flow. This may seem a little mysterious, but the forward-bias electric field is felt across this thin insulating layer without the accompanying current. This means that the I_D vs. V_{DS} characteristic curves will have response curves above $V_{GS} = 0$, as shown in Figure 6-8. We recall that the junction FET operates with reverse bias and relies on depleted carriers from the channel. Because of its characteristics the MOSFET relies on reverse bias, and there-

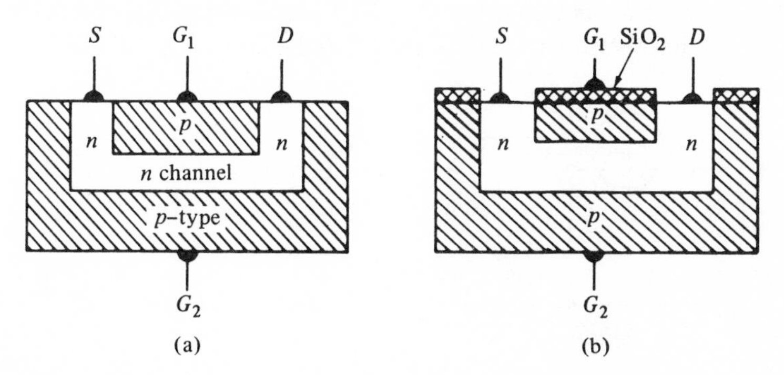

Figure 6-7 (a) Simplified diagram of junction FET. (b) Simplified diagram of insulated gate FET.

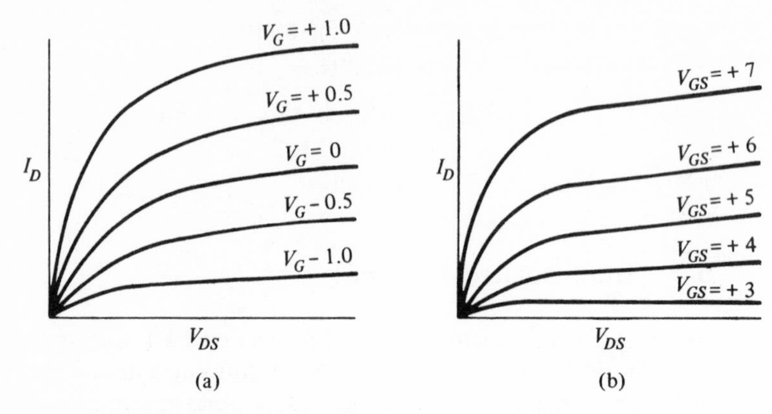

Figure 6–8 Drain family of curves for (a) enchancement–depletion type of MOSFET and (b) enchancement-type MOSFET.

fore depletion; moreover, it employs forward bias (above $V_{GS} = 0$) and "enhances" carriers into the channel. Frequently, this type of MOSFET is described as being in the depletion–enhancement mode of operation. One can readily see that interchanging JFETs and MOSFETs in an operating circuit requires investigation of the operating point.

Finally, if the *n* channel is removed in Figure 6-7(b), current will still flow between source and drain if a finite forward bias is placed between the gate and source. What essentially happens is that negative charges are induced near the gate, and these charges are swept into the source-to-drain circuit. However, the current will not flow until a finite forward bias appears. This bias is appropriately called the threshold voltage. A MOSFET of this type is called an enhancement device; its characteristics are shown in Figure 6-8(b). Notice that drain current does not flow until a certain forward bias is reached. This type definitely cannot be substituted in JFET circuits because the regions of operation are not identical.

Two typical circuits employing MOSFET are shown in Figure 6-9. One can readily see that they are not unlike the CS and CD circuits for JFET. Indeed,

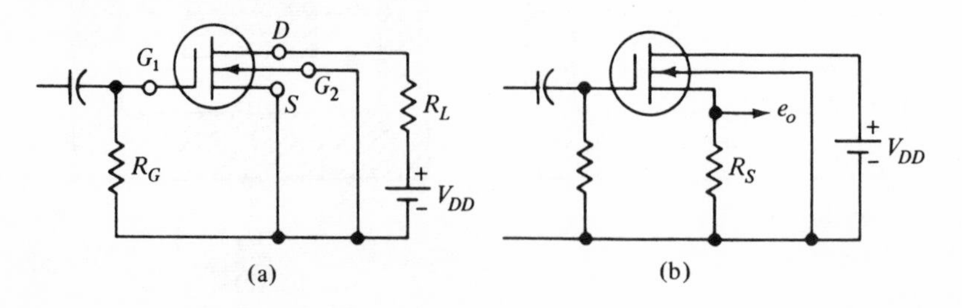

Figure 6–9 Typical MOSFET (a) common-source amplifier and (b) common-drain amplifier.

the JFET voltage equations are appropriate. Notice that the insulated gate is indicated by the spacing between the gate and the remaining portion of the symbol. Gate 2 indicates the type of doping. In this case, the gates are of the p type and therefore the MOSFET is an n-channel type.

MOSFETs usually are transported with all four leads short-circuited to minimize the chances that static electricity will puncture the thin SiO_2 layer. Hence, the student is cautioned to use the same precautions in handling until the FET is ready for use.

6-8 Summary

In this chapter it became apparent that the field-effect transistor is quite similar to the ordinary transistor. The three FET elements—source, drain, and gate—are analogous to the emitter, collector, and base, respectively. One operational difference is that the FET is a voltage-controlled device, whereas the transistor is a current-controlled device.

We also noticed that output characteristics look quite similar in cases where both are used to simulate constant current sources. In transistors the parameter that connected output and input is beta (or h_{fe}). In FETs this marriage is achieved through transconductance, which is a relationship between output current and input voltage. This relationship is used to determine voltage gain, which is $g_m R_L$ in a simple common-source amplifier and $g_m R_S/(1 + g_m R_S)$ for a common-drain (source-follower) amplifier.

Because of variations in FET parameters within the same type, it was found desirable to measure I_{DSS} and g_{m0} so that exact values of g_m could be used in circuit designs. I_{DSS} is the drain-to-source current with the gate short-circuited and g_{m0} is the transconductance at $V_{GS} = 0$. Subsequent formulas permit us to obtain the pinch-off voltage, and then g_m, at any operating V_{GS}.

Characteristically, the CS, just like the CE, produces a gain greater than one and yields phase inversion. The source follower, like the emitter follower, produces a gain less than one and yields no phase inversion. One distinct difference between FETs and ordinary transistors is the range of input resistance for these configurations. Putting it simply, the FET moves this resistance into or near the megohm range.

Finally, an insulating layer between the gate and channel produces a metal oxide semiconductor FET. One type of MOSFET permits forward- and reverse-gate bias operation. Another type functions only when forward bias is applied. The former is called enhancement–depletion mode and the latter is called enhancement. In either case the insulating layer raises the input resistance relative to the JFET.

Questions and Problems

1. Describe how (a) current flow is established and (b) current is controlled in a junction FET.
2. What are the three FET elements equivalent to in an ordinary transistor?
3. Define pinch-off voltage.
4. In Figure 6-2, determine g_m near the point (a) $V_{DS} = 4.0$ and $V_{GS} = -2.0$ volts, (b) $V_{DS} = 4.0$ and $V_{GS} = -1.5$, (c) $V_{DS} = 4$ and $V_{GS} = -1.0$, (d) $V_{DS} = 4.0$ and $V_{GS} = 0$.
5. Plot the preceding data as g_m versus V_{GS}.
6. Define I_{DSS}.
7. With the aid of a diagram, explain how I_{DSS} can be obtained in the laboratory.
8. Sketch a *p*-channel FET with two proper dc voltages. Does raising the gate-to-source voltage increase or decrease the drain current?
9. The following test data were taken in the laboratory on an *n*-channel FET:

Test	V_{GS}	V_{DS}	I_D
1	0	3.5 V	10.0 mA
2	−0.25 V	3.5 V	9.0 mA

Determine (a) I_{DSS} and (b) g_{m0}.

10. In Figure 6-4, $V_{GG} = 4.5$ volts, $R_G = 1.0\ \mathrm{M\Omega}$, $R_L = 5.0\ \mathrm{k\Omega}$, and $V_{DS} = 3.5$ volts. If the FET has 100×10^{-9} amperes gate–source leakage current, what is the voltage between gate and source?
11. In Problem 10, R_G is raised to 10 MΩ. What is the new V_{GS}?
12. In Problem 10, assume $I_D = 4.0$ mA. What value of V_{DD} is required?
13. A circuit as shown in Figure 6-4 is used. If $R_L = 5\ \mathrm{k\Omega}$ and $V_{DD} = 20$ volts, what is the approximate maximum current that can flow in the drain circuit? What is a reasonable operating drain current?
14. Determine the voltage gain if an FET uses a load resistance of 20 kΩ and is operated at a point of $V_{GS} = 0$ in Problem 9.
15. Tests on an FET produced the following:

V_{GS}, *volts*	g_m, *micromhos*
0	2000
−2	1000
−4	500

Determine A_v for each condition if $R_L = 20$ kΩ.

16. A test on an FET produced an I_{DSS} of 5.0 mA and a g_{m0} of 4000 μmho. Determine (a) V_{PO}, (b) g_m at $V_{GS} = 1.0$ volt, (c) voltage gain at $V_{GS} = 1.0$ if $R_L = 20$ kΩ in a CS amplifier.
17. For Problem 9, determine (a) V_{PO} and (b) A_v if $R_L = 5.0$ kΩ and $V_{GS} = -2.5$ volts.
18. The FET in Problem 15 feeds an ordinary transistor through a low-reactance capacitor. If the effective input resistance to this transistor is 1.0 kΩ, redetermine the three gains of Problem 15. What kind of transistor stage should follow the FET to minimize the reducing effect on the FET gain?
19. A common-drain amplifier of Figure 6-5 has the following values: $V_{GG} = 4.0$, $R_G = 1.0$ MΩ, $R_S = 1.0$ kΩ, $V_{DD} = 9$ volts. If the gate–source leakage current is 1.0 μA, $I_D = 2.0$ mA, and $g_m = 1000$ μmho, determine (a) V_{GS}, (b) V_{DS}, (c) A_v, and (d) R_o.
20. Describe the construction of an insulated-gate FET.
21. What are some characteristics of an IGFET?
22. What does MOSFET mean?

7

The Unijunction Transistor (UJT)

7-1 Basic Operation

The unijunction transistor is a three-terminal device, as is the ordinary transistor; as a matter of fact, its physical resemblance to the transistor is quite close. However, the applications are different. Ordinary transistors are used in amplifiers and in many other types of circuits. The UJT, on the other hand, cannot amplify; however, it is used in oscillators and in timing and multivibrator circuits. Although transistors can perform these functions, the UJT requires less circuiting and, thus, simpler designs.

Figure 7-1(a) shows the basic components of a UJT. Ohmic contacts are made to ends of an n-doped silicon lead. These are identified as base 1 and base 2. On one side a p region, called the emitter, is made; it is physically closer to B_2 than to B_1. (One can see the reason for a nomenclature of

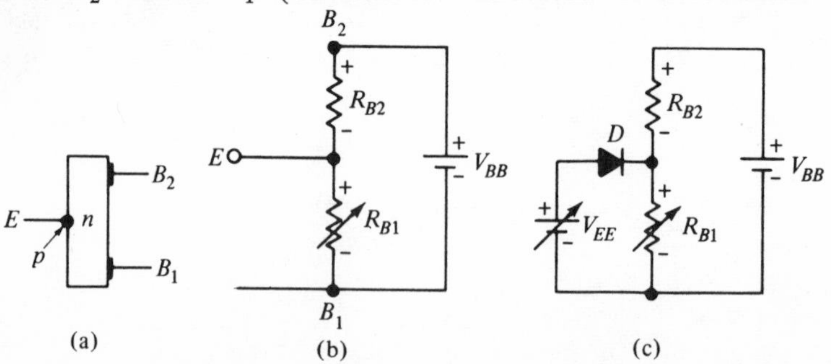

Figure 7-1 (a) Pictorial representation of a UJT. (b) Equivalent circuit of a UJT. (c) Equivalent circuit and dc potentials on a UJT.

unijunction or double-base transistor.) B_2 is made positive in potential with respect to B_1 and the emitter is forward-biased with respect to B_1. The uniqueness of a UJT relies on the amount of forward bias between emitter and base 1. To further clarify this point, Figure 7-1(b) shows an equivalent circuit of the *n* material with V_{BB} applied and no forward bias at the emitter. Essentially, we have a voltage divider where the voltage that appears between the emitter and base 1 depends on the magnitude of R_{B1} relative to R_{B2}. Incidentally, the sum of R_{B1} and R_{B2} ranges between 4 kΩ and 10 kΩ and R_{B2} is less than R_{B1} because B_2 is closer to the emitter lead. Similar to voltage-divider circuits that are delivering zero load current, an increase in V_{BB} increases the voltage between the emitter and base 1. Thus the base-to-base circuit is, in essence, a voltage-divider circuit, and the voltage that appears across R_{B1} is evidently important. The magnitude of this voltage is

$$V_{R_{B1}} = V_{BB} \frac{R_{B1}}{R_{B1} + R_{B2}} \tag{7-1}$$

We can now proceed a little further by impressing forward bias in the equivalent circuit as shown in Figure 7-1(c). Notice the addition of an equivalent diode *D* in the circuit. If we sum up the voltage across diode *D*, we obtain

$$V_{EE} - V_{R_{B1}} = V_D \tag{7-2}$$

This tells us that the bias for the diode can be in the forward or reverse direction depending on the relative magnitude of the two voltages, V_{EE} and $V_{R_{B2}}$. The polarity signs on the diagram tells us that V_{EE} is trying to forward bias *D*, and $V_{R_{B2}}$ reverse bias. We also know that $V_{R_{B2}}$ depends on V_{BB}. Let us fix the value of V_{BB} and hence $V_{R_{B2}}$ and observe the action as we proceed from V_{EE} equals zero to V_{EE} greater than $V_{R_{B2}}$. For all intents and purposes, this indeed is the important phenomenon of a UJT. If V_{EE} is zero, $V_{R_{B2}}$ predominates, *D* is reverse-biased and no current flows (only minute reverse current flows) in the emitter lead. If we increase V_{EE} to a point where it is greater than $V_{R_{B2}}$, the diode is forward-biased. The peculiar feature is that the increase in current is not gradual, but quite abrupt. This action can be described as follows: When the *p* material becomes positive, holes are injected toward the negative terminal or base 1. A movement of carriers, holes, implies less resistance (or a decrease in R_{B2}). A decrease in R_{B2} means that the percent of V_{BB} that R_{B2} takes is decreased. If $V_{R_{B2}}$ is decreased (reverse bias) then the forward bias V_{EE} becomes more effective and causes more emitter current to flow; see Equation (7-2). What follows next is another idiocyncracy of the UJT. As the current in the emitter increases, the voltage between the emitter and base 1 *decreases* in an uncontrollable fashion. One normally associates an increase in current with an increase in voltage, but in this case the opposite is true. This type of characteristic is exhibited by a few electronic devices and the phenomenon is called *negative resistance*. Once the UJT goes through this unstable region, it returns to a stable behavior; that is, an increase in I_E corresponds to an increase in V_{EB1}. It is this negative resistance characteristic that makes the UJT a useful device in oscillators and timing circuits.

To solidify the action of a UJT, a short review is appropriate. A unijunction transistor has resistance between its bases. This divider action caused by the base resistances tends to reverse bias the emitter-to-base-1 junction. Once external forward bias overcomes this internally developed reverse bias, emitter current increases rapidly. Then the emitter–base-1 voltage decreases with increasing current in an uncontrolled manner. Once through this region, called negative resistance, the UJT behaves in a normal manner; that is, if voltage increases, the current increases.

Before we leave this behavioral explanation, it is worth noting what effect an increasing V_{BB} might have on the V_E-I_E characteristic. Simply, a larger V_{BB} means $V_{R_{B2}}$ is larger. If $V_{R_{B2}}$ is larger, then the forward bias V_{EE} will have to go to a larger value before the UJT can "skate" through its characteristic response. Putting it another way, we can expect a family of curves of V_E vs. I_E for different values of V_{BB}.

7-2 Characteristic Curves and Specifications

A better summary for the behavior of an electronic device is the static characteristic curve. Figure 7-2(a) shows a smoothed-out and approximate I_E-V_E characteristic of a UJT. Note that any current to the left of the vertical axis indicates reverse current flow. Hence, from point *a* to *b* we have reverse current flow because $V_{R_{B2}}$ dominates the net bias voltage. Once we get to the right of the vertical axis V_{EE} (forward bias) predominates and forward current rises from *b* to *c*. The current continues to increase as the net voltage from emitter to base 1 slides downward or decreases through its negative-resistance region. This is the region from *c* to *d*. Finally, the UJT returns to a positive resistance behavior from *d* to *e*.

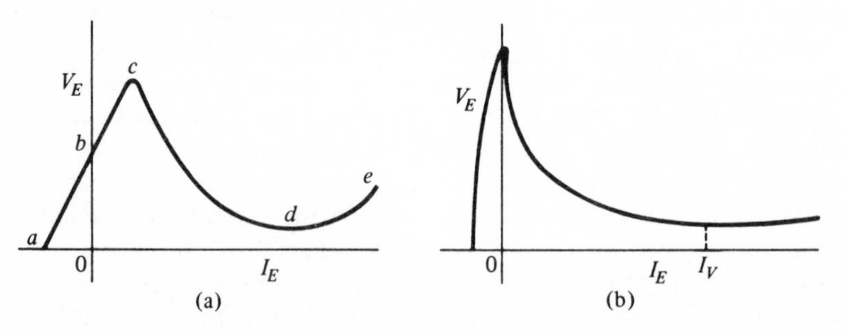

Figure 7-2 (a) Approximate $V_E - I_E$ characteristic of a UJT. (b) More representative $V_E - I_E$ characteristic of a UJT.

This response is appropriate for a particular V_{BB}. A similar curve would result if V_{BB} were raised except that the peak value (point *c*) would reach a higher value. The remaining portion, *cde*, would follow except with higher V_E values.

A more accurate V_E-I_E characteristic curve is shown in Figure 7-2(c). Notice that there is more curvature and that the peak voltage occurs practically at zero emitter current. This means that the negative-resistance region starts practically at $I_E = 0$.

At this point it is worth our while to collect and explain some of the terminology used with UJT. The magnitudes associated with these terms are specifications and used frequently in UJT circuit design. Some of the UJT's specifications can be related to Figure 7-2(a). These are enumerated below:

1. I_p, the peak-point current, is the emitter current at the peak point. This is point *c* on Figure 7-2(a). Manufacturers quote this value for a particular V_{BB} and temperature.
2. V_p, the peak-point emitter voltage, is the corresponding emitter voltage for peak-point emitter current. It, too, is obtained at point *c* in Figure 7-2(a).
3. I_v, the valley current, is this emitter current at the valley point, which is *d* in Figure 7-2(a).
4. V_v, the valley voltage, is the emitter voltage at the valley point which is point *d* in Figure 7-2(a). The valley voltage depends on V_{BB} so manufacturers quote a range of valley voltages expected or display a family of curves that reveal the valley voltage.

One parameter that is quoted and used in circuit equations is η (Greek letter "eta"). It is called the intrinsic standoff ratio. The expression "standoff" assists us in understanding the meaning of η. It is defined as the ratio

$$\eta = \frac{R_{B1}}{R_{B1} + R_{B2}} \tag{7-3}$$

This ratio was used in Equation (7-1), which yielded the proportion of reverse-bias voltage that can be expected intrinsically from V_{BB}. Repeating the equation,

$$V_{R_{B2}} = V_{BB} \frac{R_{B1}}{R_{B1} + R_{B2}} \tag{7-4}$$

or

$$V_{R_{B2}} = \eta V_{BB} \tag{7-5}$$

Finally, we can see that as long as the forward bias in the emitter lead is equal to or less than $V_{R_{B2}}$ we can expect a stand off. Or putting it another way, the UJT will not have any important action as long as the reverse bias predominates. The amount of this reverse bias is ηV_{BB}. A typical value of η is 0.56.

If we pursue exploring η, we can mathematically express and connect it to the peak voltage V_p, which essentially is the critical voltage where the negative resistance starts:

$$V_p = \eta V_{BB} + V_D \tag{7-6}$$

where V_D is the forward bias required to turn on the emitter diode. In this case, V_D equals approximately 0.7 volt at room temperature. This equation simply means that the critical action at V_p will be reached if enough forward bias is applied. The amount required is ηV_{BB}, to overcome the intrinsic standoff voltage, plus a little more (which is V_D), to put the emitter sufficiently in forward bias for conduction. Once this is done, the UJT slides through its typical negative-resistance region.

Another characteristic that manufacturers quote for UJT is R_{BB}. This is the resistance from B_1 to B_2 or $R_{B1} + R_{B2}$. A typical value is 5.6 kΩ. This quantity is quoted at a specific temperature because it is temperature-sensitive. The standoff ratio η, on the other hand, is quite constant over a wide range of temperature. One can expect this because η is a ratio of resistances and, if the resistances are affected equally by temperature change, their ratios should remain unchanged.

Typically, for semiconductor devices, maximum values of current, voltage, and power dissipation are also quoted by manufacturers of UJTs. The emitter current must be kept below the specified peak value. Limitations for maximum emitter reverse voltage and maximum interbase voltage must also be observed. The maximum power dissipation refers to the maximum permissible power dissipated by the emitter and interbase resistance.

Indeed, all these limitations and specifications may seem overwhelming. It can be said that all solid-state devices have more than their share of specifications. Only with frequent experience can one become sufficiently knowledgeable. Generally, most devices have maximum ratings that are self-explanatory, such as maximum collector current for a transistor. Along with these are parameters that are characteristic for the device. Examples of these are h_{fe} for transistors and g_{m0} for FETs. Finally, a host of minor characteristics follow that sometimes may confuse the picture. One of these is the reverse-current leakage in an ordinary diode. Under some special circumstances it may be important, but it is usually ignored. In any case, all of these are generally quoted and hence may create some confusion.

EXAMPLE 1

In Figure 7-1(c), $R_{B1} = 5.0$ kΩ, $R_{B2} = 4.0$ kΩ, $V_{BB} = 18$ volts, and $V_{EE} = 5.0$ volts. Is the UJT forward-biased?

SOLUTION

$$V_{R_{B1}} = V_{BB}\frac{R_{B1}}{R_{B1}+R_{B2}}$$
$$= 18\frac{5(10^3)}{(5+4)(10^3)}$$
$$= 10 \text{ volts}$$

Since $V_{R_{B1}}$ is greater than V_{EE}, the UJT is reverse-biased.

EXAMPLE 2

In Example 1, what value of V_{EE} will cause the UJT to fire?

SOLUTION

Since the diode has to be forward-biased by approximately 0.7 volt,

$$V_p = \eta V_{BB} + V_D$$
$$= \frac{R_{B1}}{R_{B1}+R_{B2}} V_{BB} + V_D$$
$$= \frac{5(10^3)}{9(10^3)} 18 + 0.7$$
$$= 10.7 \text{ volts}$$

7-3 Relaxation Oscillator

There is a limit to the number of words that may adequately describe the behavior of any electronic device. Generally, an appropriate application helps to "clear the air." The UJT is no exception. The example that follows should yield more insight to its behavior. One word of caution, however. The UJT cannot be used in an amplifier stage. Instead, the UJT is a special device, and essentially it can produce a pulse or periodic pulses when conditions are right at the emitter. Hence, whenever short duration pulses are required in a circuit the UJT with its associated circuit can do the job.

Figure 7-3(a) shows a relaxation oscillator that employs a unijunction transistor. Notice the symbol for the UJT, which is quite similar to that of the FET. As anticipated, a potential is applied across the two bases with B_2 positive with respect to B_1. The output, which is across R_2, depends on what happens between the emitter and B_1. From the diagram it is apparent that the voltage across the capacitor C will affect this emitter–base-1 diode.

As a short digression, let us look at R_1, C, and V_{BB}. This circuit is a typical RC circuit in which the voltage across C will attempt to reach the value of

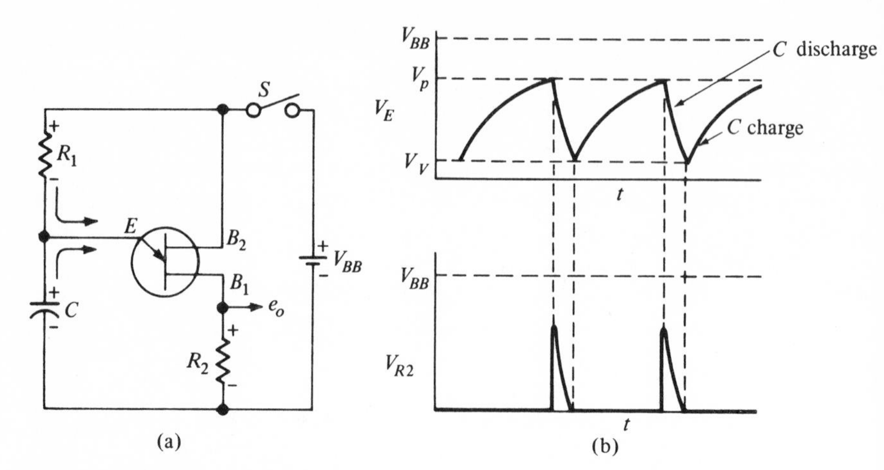

Figure 7–3 (a) A UJT relaxation oscillator. (b) Emitter-voltage waveform. (c) Output-voltage waveform.

V_{BB} once switch S is closed. The time it takes for C to charge to a specific value depends on the values of R_1, C, and V_{BB}. Therefore, once the switch is closed the capacitor starts to charge, its voltage approaching V_{BB}. While this is going on, the UJT is somewhat idle because the forward bias on the emitter diode is not sufficient to cause any current flow. As the capacitor continues to charge to the polarity indicated on the diagram, it becomes more likely that the emitter diode will become forward-biased sufficiently to trigger the UJT through its unstable region. We briefly pause in describing this action to mention that there is an apparent requirement that V_{BB}–I_1R_1 be high enough to reach or go beyond the V_p of the transistor. Recall that once V_p is reached the UJT loses control and more current flows in the emitter lead at a lower voltage. If we cannot guarantee that the voltage across the capacitor will reach V_p then no action will occur.

Let us assume that the capacitor will eventually reach V_p. At this moment, the base-1 resistance decreases rapidly. This in turn causes the capacitor to discharge quite rapidly through the emitter–base-1 junction and R_2. Because of this new low-resistance path, V_{BB} also supplies current to the emitter by way of R_1. Now we see that V_{BB} and capacitor C are supplying current. However, the capacitor current tapers off and approaches zero quite quickly. This is typical of a capacitor discharging into a low resistance. Because of low capacitor current and because the current from V_{BB} through R_1 is designed not to go beyond the valley point, the UJT rebuilds its emitter–base-1 resistance to its original large value rather quickly. This means that R_1 is chosen so that the UJT operates up to but not beyond its valley point. Obviously, this is another design criterion. Once original conditions reappear, the capacitor starts to charge again, builds up to V_p, the UJT "fires," the capacitor discharges quickly, and the UJT returns to its original state. The time required for one cycle of this behavior is called a period. The reciprocal of the period is the frequency.

Figure 7-3(b) is a pictorial view of the previously described action at the emitter. Notice that the limits of the wave are V_v and V_p. Figure 7-3(c) shows the output pulse that can be used to trigger another device such as a silicon controlled rectifier or to trigger a sweep circuit in an oscilloscope. The repetition rate depends on R_1, C, R_{B2}, R_2, and V_{BB}. As a quick approximation, if we take the reciprocal of the off time of the UJT, we can extract the frequency of the pulses produced. This of course requires observing the off time on a horizontally calibrated oscilloscope. Hence,

$$f \simeq \frac{1}{t_{\text{off}}} \tag{7-7}$$

where t_{off} is the time when the UJT is not conduction or, in other words, the time required for the capacitor to charge from V_v to V_p.

To ensure all this action and hence an output, the equations that describe the implied design criteria from the previous paragraph are

$$\frac{V_{BB} - V_p}{R_1} > I_p \tag{7-8}$$

and

$$\frac{V_{BB} - V_v}{R_1} < I_v \tag{7-9}$$

Satisfying Equation (7-8) ensures that the UJT will "fire" or start conducting. Putting it another way, R_1 and V_{BB} are chosen so that sufficient voltage remains to reach V_p. Equation (7-9) guarantees that the UJT does not operate beyond the valley point. This apparently means that we are trying to operate in the unstable region. Since this is impossible, the UJT returns to its only stable operating point: $I_E \simeq 0$. Keep in mind that a UJT has two stable operating regions. Referring to Figure 7-2(a), they are, from *a* to *c* and from *d* to *e*. Since R_1 is chosen to prohibit operation from *d* to *e*, the UJT has no choice but to return to the *a*-to-*c* region should it find itself momentarily in the unstable region between *c* and *d*. This can be more dramatically demonstrated by Figure 7-4(a).

The characteristic represents a V_E-I_E response for a UJT at a particular V_{BB}. If R_1 is plotted so that it intercepts the stable region between *a* and *c* and the unstable region to the left of I_v, line *A* would result. The slope of this line represents R_1. It is obvious that many lines can be drawn to satisfy the foregoing condition. Thus R_1 can assume values that range from 2 kΩ to 2 MΩ.

On the other hand, in some circuits R_1 is chosen to intercept both stable regions and a different behavior is noted; see Figure 7-4(b). If a sufficiently large pulse carries point P_1 beyond the V_p point, the UJT will "run through"

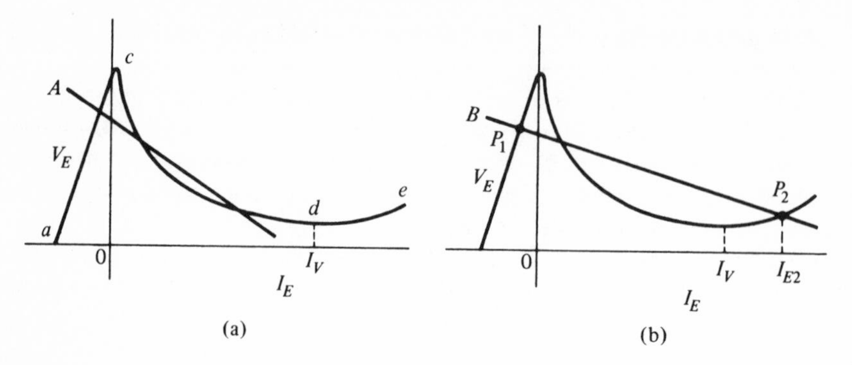

Figure 7–4 UJT characteristic (a) with *A* resistor intercepting one stable point and (b) with *B* resistor intercepting two stable points.

the unstable region and stop at point P_2. The operating UJT will remain here, supplying I_{E_2} until another but opposite pulse is deliberately injected to return the UJT to P_1. This type of behavior opens the possibilities of applying the UJT in control circuits.

EXAMPLE 3

The charging time in Figure 7-3(b) is 100 microseconds. (a) What is the approximate frequency? (b) If the discharge time is 10 μs, what is the precise frequency?

SOLUTION

(a)
$$f \simeq \frac{1}{t_{\text{off}}}$$
$$= \frac{1}{100(10^{-6})} = 10 \text{ kHz}$$

(b)
$$f = \frac{1}{T}$$
$$f = \frac{1}{(100 + 10)(10^{-6})} = 9.1 \text{ kHz}$$

EXAMPLE 4

A UJT is chosen that has the following characteristics: $V_{BB} = 30$, $I_p = 5\ \mu\text{A}$, $\eta = 0.60$, $V_v = 3.0$, and $I_v = 20$ mA. It is to be used in a relaxation oscillator. (a) Determine the approximate V_p. (b) Determine whether the two currents requirements are met for $R_1 = 1.0\ \text{M}\Omega$.

SOLUTION

(a)
$$\begin{aligned} V_p &= \eta V_{BB} + V_D \\ &= (0.6)(30) + 0.7 \\ &= 18.7 \text{ volts} \end{aligned}$$

(b)
$$\frac{V_{BB} - V_p}{R_1} > I_p$$

$$\frac{30 - 18.7}{1 \times 10^6} > I_p$$

$$11.3(10^{-6}) > I_p \qquad \text{where } I_p = 5 \times 10^{-6}$$

$$\frac{V_{BB} - V_v}{R_1} < I_v$$

$$\frac{30 - 3.0}{1 \times 10^6} < I_v$$

$$27(10^{-6}) < I_v \qquad \text{where } I_v = 20 \text{ mA}$$

Both current conditions are satisfied; therefore, the circuit will oscillate.

7-4 How a UJT Is Used in a Control Circuit

The intent of this text is to acquaint the student with most of the electronic devices in use today. As we proceed through the text, it will be obvious that these devices are many and varied in application. This of course may lead to confusion unless the student is occasionally given some practical use to whet his appetite. If some purpose for learning becomes apparent, the learning process becomes a little easier.

Figure 7-5 is introduced to serve this purpose relative to the UJT. It is a block diagram in which each block is functional. Many complicated electronic systems—television sets, for example—begin an explanation with a block diagram to give the reader an overview and a meaningful picture. Later on, the blocks or circuits are taken one at a time and explained in

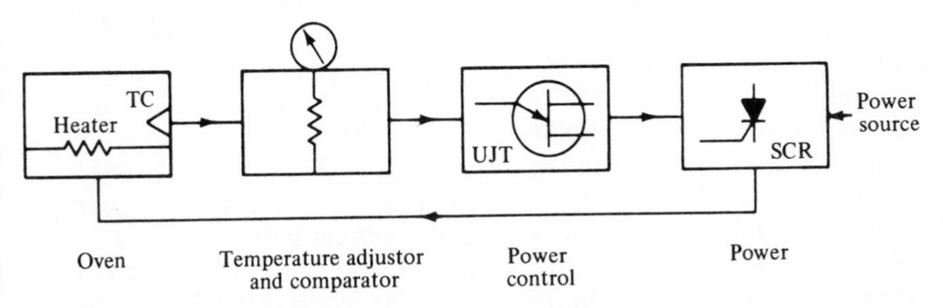

Figure 7–5 Temperature-control system.

greater detail. Figure 7-5 is an elementary block diagram depicting a temperature control system. To be more specific, the purpose of the electronic circuitry is to maintain temperature in an oven. The oven contains a heater, and the current through it must be controlled. A thermocouple (TC) produces millivolts for an output voltage. The output voltage increases as temperature increases. This output is sent to the comparator, which essentially compares the TC millivolts and the millivolts developed by the "temperature-adjust" control. If the temperature-adjust control produces a larger voltage than the TC, this means that the oven has to be heated. (This is analogous to turning up a thermostat in the home.) The polarity of this "compared" voltage is to turn the UJT on. At this point we are aware how this is done in a UJT. The output voltage of the UJT has a polarity that will fire the silicon controlled rectifier (SCR). The study of SCRs will follow, but for the moment it is sufficient to say that it is capable of handling current in the ampere range once it is turned on. The current that goes through the SCR also goes through the heater. This of course heats the oven. As the oven temperature rises, the TC millivolts increase. When these TC millivolts equal the temperature-adjust (thermostat) millivolts, no output arrives at the UJT and hence it will not conduct, and thus the SCR will not fire. If the SCR does not conduct, no current flows through the heater and hence the oven must be at the temperature designated by the temperature-adjust control.

The above is a simple illustration of how a single UJT can be part of a system. If one were troubleshooting this kind of circuit it would be necessary to know how a UJT functions and what can be expected of it under certain circumstances. It is hoped that the previous paragraphs give us sufficient insight to do just that.

7-5 Summary

This chapter revealed that a unijunction transistor is unlike an ordinary transistor. It does not amplify but produces an output only when certain conditions are satisfied at the input.

Basically, the UJT has three leads: emitter, base 1, and base 2. A dc voltage across B_2 and B_1, with B_2 positive, reflects a percentage of this voltage at the emitter. This implies that B_2–B_1 acts like a voltage divider. Hence, a larger base-to-base voltage V_{BB} reflects a larger voltage at the emitter. This reflected voltage keeps the emitter reverse-biased and therefore keeps low current flow in the emitter–base-1 circuit. If sufficient external voltage is applied to the emitter, it overcomes the intrinsic reverse bias and the emitter–base-1 current increases rapidly. This higher current flow is accompanied by a lower voltage. This phenomenon is called negative resistance. The UJT cannot maintain a current flow in this region, which therefore is called an unstable region. The amount of external voltage necessary to trigger the UJT is at least $\eta V_{BB} + V_D$, where η is the standoff ratio.

One use of the UJT is in a relaxation oscillator. In this circuit an RC circuit is charged by a dc potential. The capacitor voltage rises from zero, and when it reaches $\eta V_{BB} + V_D$ the UJT fires. The capacitor discharges and the process starts all over again. This procedure produces periodic pulses at a frequency that is determined by the value of V_{BB}, R, and C.

Questions and Problems

1. Sketch the placement of the three elements in a UJT. How are resistances R_{B1} and R_{B2} related?
2. Sketch the equivalent circuit of a UJT.
3. With the use of the equivalent circuit, explain how the UJT fires.
4. If $R_{B1} = 6$ kΩ, $R_{B2} = 4$ kΩ, and $V_{BB} = 24$ volts, what value of voltage appears as reverse bias in the UJT?
5. Repeat Problem 4 for $V_{BB} = 10$ volts.
6. What is η in Problem 4?
7. If $V_D = 0.7$ volt, what is V_p in Problem 4?
8. Why is V_p important?
9. In Figure 7-1(c), $R_{B1} = 5.0$ kΩ, $R_{B2} = 5.0$ kΩ, $V_{BB} = 12$ volts, and $V_{EE} = 5.0$ volts. Is the UJT forward-biased?
10. Describe the action in Figure 7-3.
11. The UJT of Problem 9 is used in a relaxation oscillator as shown in Figure 7-3(a). If $V_{BB} = 12$, $I_p = 4\ \mu$A, $V_v = 4.0$, $I_v = 10$ mA, $R_1 = 50$ kΩ, and $C = .01\ \mu$F, determine (a) η and (b) whether the oscillator conditions are satisfied.
12. If the charge time is 0.4 ms in Problem 11, determine the approximate frequency.
13. Sketch a typical V_E-I_E UJT characteristic curve.
14. Repeat Problem 13 for the case in which V_{BB_2} is larger than V_{BB_1}.

8

The Silicon Controlled Rectifier (SCR)

8-1 Structure and Basic Operation

The silicon controlled rectifier is another special type of device for use in specific applications. It is not used as an amplifier. Its operation is quite like the unijunction transistor in that nothing occurs unless certain conditions are met; then a large amount of current flows. The current-handling ability is much greater than that of the UJT; moreover, the basic structure is quite different.

Figure 8-1(a) shows a four-layer semiconductor sandwich. The outermost *p* layer is called the anode, the outermost *n* layer is the cathode, and the

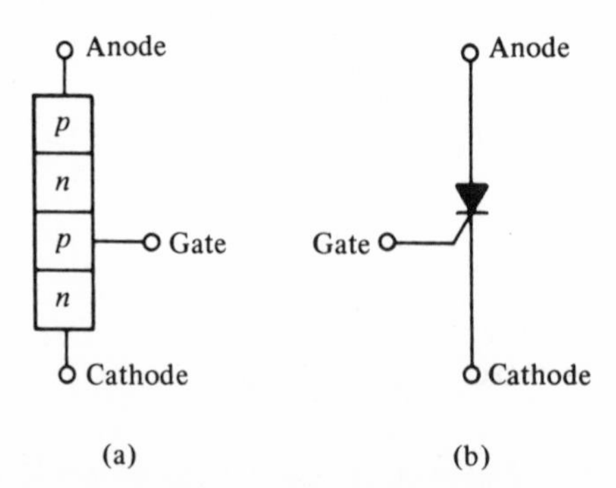

Figure 8–1 The SCR (a) construction diagram and (b) symbol.

inner *p* is the gate. This structure constitutes an SCR. Figure 8-1(b) shows the symbol. The word *gate* once again implies that it has control of whatever happens between the remaining two elements. The remaining two parts are anode and cathode and they in turn imply rectification. Hence the purpose and meaning of the silicon *controlled rectifier* are understood.

If an anode is made positive with respect to a cathode in a given ordinary diode, one would expect current to flow because we would have a condition of forward bias. The current would be limited by the amount of external resistance. If we used the same approach on the SCR—that is, placing a positive potential on the anode with respect to the cathode—current *will not* flow. In the SCR's case we have three diodes in series; Figure 8-2 shows that diode 2 is reverse-biased. One can envision that V_{BB} is distributed among three resistors that are equivalent to forward-biased resistance for D_1 and D_3 and reverse-biased equivalent resistance for D_2. Therefore, the current flow in this circuit will be determined by the diode leakage current of D_2. This of course is a small quantity and hence a high resistance. Putting it another way, the SCR is in the "off" state because current flow is very small.

To change this state it is obvious that the gate must be called upon to do its part. To understand this operation a full equivalent circuit must be used as shown in Figure 8-3. Keep in mind that the purpose of an equivalent circuit is to simulate a given device with less complicated components to make it more understandable. Notice that one of the two transistors is *n-p-n*; the other is *p-n-p*. It is readily seen in Figure 8-3(a) that base 1 is common to collector 2, and collector 1 is common to base 2. This is seen schematically in Figure 8-3(b).

We can duplicate our path of Figure 8-2 where we realize low current will flow because one diode is reverse-biased. In Figure 8-3(b) the distribution of V_{BB} is such that the emitter–base diode of T_1 is forward-biased. Continuing to T_2, the collector–base diode is reverse-biased and the base–emitter diode is forward-biased. These polarities are indicated in Figure 8-3(b).

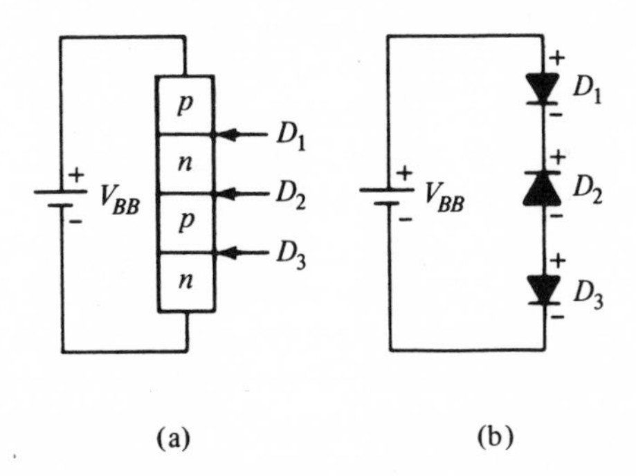

Figure 8–2 (a) SCR junctions and (b) biasing of junctions when the anode is positive with respect to cathode.

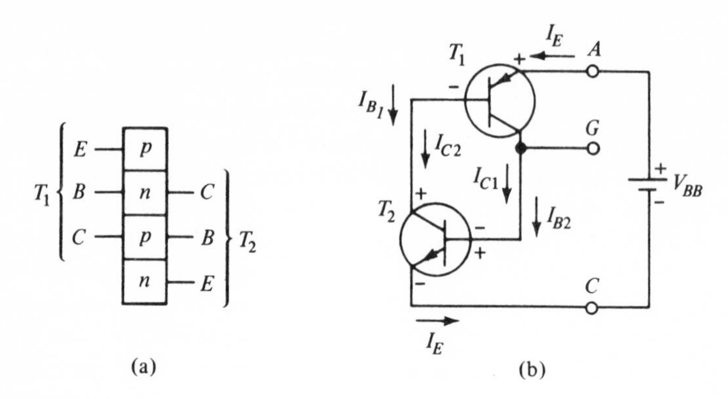

Figure 8-3 The SCR (a) Equivalent diagram. (b) Equivalent two-transistor circuit.

These latter polarities for T_2 imply that T_2 should conduct because the biases (E_2–B_2 forward and B_2–C_2 reverse) are in the proper direction. However, the voltage division according to resistance that each diode offers is such that the emitter–base diode of T_2 is not forward-biased enough to cause high emitter current and hence high collector current. If we took another path, such as E_1–B_1, to B_1–C_1, to B_2–E_2, a similar low current I_{C1} would result. These two low currents, I_{C1} and I_{C2} increase slowly as V_{BB} increases, and if we raise V_{BB} high enough we can expect large currents. But before we investigate that avenue let us assume low currents for I_{C1} and I_{C2} and observe the effect of placing a positive voltage on the gate with respect to the cathode. A positive gate voltage will forward-bias the base-to-emitter diode of T_2 a little more. This increase in forward bias causes more collector current I_{C2}. (We recall transistor theory from Chapter 4.) Since I_{C2} is also the base current for T_1, an increase in I_{B1} causes an increase in I_{C1}. I_{C1} also happens to be the base current for T_2 and therefore an increase in I_{C1} causes I_{B2} to increase. This in turn causes I_{C2} to rise, and so on. The upshot of the whole thing is that a positive pulse has caused a snowballing effect, producing large collector currents; moreover, apparently the gate has also lost control because of this regenerative effect. Finally, since I_C is related to I_E, we can conclude that the emitter current I_E jumped to a large value once the gate voltage triggered the circuit. As a matter of fact, the current is limited primarily by a load resistance placed in series with the emitter lead. This current will flow until a high negative pulse is applied to the gate or the battery supply is reduced to zero. This means that the gate will regain control if either of the two conditions is applied. Projecting ahead a little, one automatic way of reducing the battery to zero is to replace it with an ac source. The alternating current automatically reduces itself to zero every time it crosses the axis, thus permitting the gate to regain control if regaining control is required. This concept will be investigated later in this chapter.

We can conclude our investigation of the equivalent circuit by saying that high current can flow if we raise the power voltage high enough or if we inject a low forward-bias current to initiate the regeneration effect. In either case, the external current is relatively high; this conducting condition for an SCR is called the "on" state. Some SCRs are capable of delivering more than 500 amperes.

8-2 Firing Characteristics

A detailed study of an SCR characteristic at this point not only will help in understanding the behavior but will also assist in understanding some of the manufacturers' ratings. Figure 8-4(a) shows a typical anode current versus anode to cathode voltage for an SCR or for that matter the I-V characteristic of the equivalent circuit of Figure 8-3(b). The first-quadrant response (anode positive with respect to cathode) is more important than the third-quadrant response (reverse bias). If the gate current is assumed to be zero, a rise in anode voltage results in a slowly increasing anode current. At a particular point A, the SCR fires and the anode-cathode voltage drops to an approximate lower value B. The SCR is now "on" and the external current and anode–cathode voltage will settle on some operating point along B to C depending on the amount of external resistance in series with the anode. To regain control, the anode current has to be decreased to a value less than I_{ho}, which is the minimum holding current.

If a small gate current, I_{G1}, is permitted to flow and the characteristic response is then plotted, the same general response is noted except that the SCR fires at a lower anode voltage. This is seen in Figure 8-4(b). If a larger current, I_{G2}, flows, the breakover voltage is further reduced. Hence, as the gate current is increased, the breakover voltage decreases and the SCR will go into the "on" state (high current) with a lower applied anode–cathode

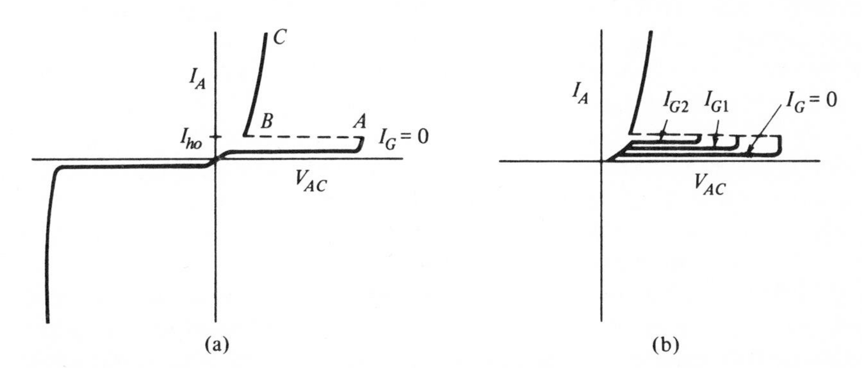

Figure 8–4 (a) SCR forward and reverse bias response. (b) SCR forward response with different gate currents.

voltage. We can conclude that the SCR can supply heavy current to a particular device by applying a small amount of gate current. This can be achieved by maintaining a fixed dc voltage between the anode and cathode and then injecting a current into the gate. The only problem is that once the SCR fires, it stays "on." There may be applications where we wish to truly *control* the amount of anode-current flow through a load. This indeed is done in many circuits by replacing the dc source with an ac source. Figure 8-5 can help visualize the control possibilities of the SCR. It must be remembered that, at any time,

$$V_{BB} = V_{RL} + V_{AC} \tag{8-1}$$

This applies to instantaneous ac as well as to dc voltages. For example, if the anode current in Figure 8-5(a) is zero, then the voltage across R_L is zero and V_{AC} equals V_{BB}. Therefore, once switch S is closed, the anode-to-cathode voltage follows V_{BB}. This is seen in the early part of Figures 8-5(b) and (c). If there is a finite I_G flowing, the rising cathode–anode voltage will respond by firing at A and then falling to B. For the remaining portion of the positive half of the source voltage the SCR will be "on." At this point, the question may be asked, "What happened to the remaining positive portion of the sine wave?" The wave shown in Figure 8-5(c) is that which appears between the cathode and anode. If Equation (8-1) is appropriate, then the remaining positive portion of the wave appears across the load resistance. These portions are shown in Figure 8-5(d). Notice that they are all on one side of the axis and the average of these portions is the direct current that flows through the load.

Now suppose I_G is increased to I_{G2}. We can expect the SCR to fire sooner along the sine wave because of the basic theory of operation. Figure 8-5(e) shows this earlier firing. One can see that the remaining sine-wave portion will be larger and therefore yield a higher dc load voltage. The amount of load voltage can be controlled by causing the SCR to fire almost immediately or at zero degrees. This rectification process will yield the maximum possible load voltage. The I_G can be reduced to the point at which conduction occurs from 90° to 180° on the source sine wave. This is the lowest controlled current. We cannot delay the firing until the positive half of the sine wave is on its way down. This means we cannot effect a control, from 90° to 180° —on the source sine wave. This is so because the SCR fires at the first firing potential. These firing potentials are duplicated on the second half (90°–180°) of the positive wave and therefore have no effect because the SCR is already fired when the first firing voltage appears.

Finally, we must make note that the negative portion of the sine wave reverse-biases the SCR. In the process of going from forward to reverse bias the anode-to-cathode potential was momentarily at zero volts, which is equivalent to an open switch; therefore, the anode-to-cathode circuit is momentarily opened and the gate is permitted to regain control. This auto-

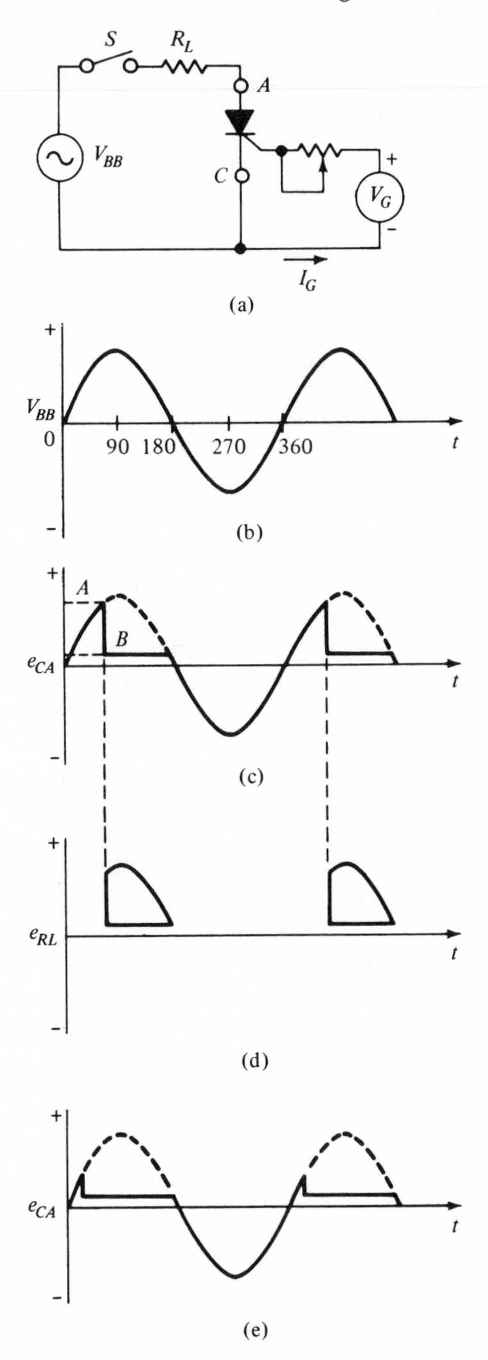

Figure 8–5 (a) SCR circuit with ac voltage between anode and cathode. (b) Supply-voltage waveform. (c) Cathode-to-anode voltage. (d) Load voltage. (c) Cathode-to-anode voltage with less gate current.

matic switching capability shows why alternating current is used in many operating circuits.

8-3 Ratings and Characteristics

Once again we find it necessary to discriminate between devices by observing the ratings and the terminology that goes with the ratings. There is indeed more than the desired amount of identifying characteristics. To minimize confusion, only the more obvious and important parameters are discussed.

As a start, a typical characteristic is shown in Figure 8-6. From it, anode characteristics can be identified.

I_{ho}, the holding current, is the minimum current that must flow through the SCR in the forward direction to ensure that it will remain in the conducting state if the gate signal is removed. See Figure 8-6.

V_{rb}, the peak reverse voltage, is the maximum peak voltage that can be applied between anode and cathode, without the device going into an inverse avalanche breakdown. This is a typical rating for ordinary diodes. Figure 8-6 shows it is a value below the avalanche (nondesirable for operation) region.

V_{fb}, the peak forward blocking voltage, is the maximum peak forward voltage that can be applied between anode and cathode and still have the SCR in the "off" state. This parameter is important because in a given application the anode–cathode voltage is usually present at all times and the firing is controlled by the gate. With no gate current the SCR is usually not fired.

V_f, the forward voltage drop, is the voltage between the anode and cathode when the SCR is conducting. It, of course, depends on how much current is

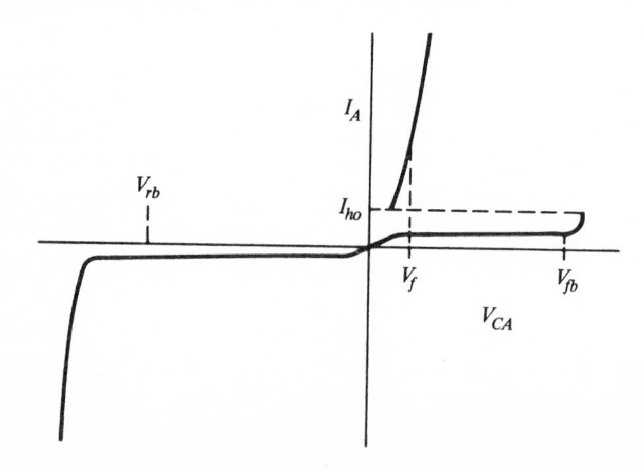

Figure 8–6 SCR anode characteristics with identifying ratings.

flowing through the SCR, and in Figure 8-6 it is any place along the conducting portion of the characteristic.

I_{avg}, the average forward current, is the maximum average current that the SCR can conduct under certain conditions. The manufacturers usually supply curves that derate the maximum average current as SCR temperature is increased. Each application must be investigated so that the selected SCR has the capability to meet this current requirement.

The following parameters deal with the gate–cathode ratings:

I_{gt}, the minimum gate current, is the minimum gate trigger current that will cause the SCR to switch "on" for a given anode–cathode voltage.

I_{gfm}, the peak forward gate current, is the maximum peak current that can flow through the gate–cathode junction without causing damage.

V_{gt}, the gate trigger voltage, is the gate–cathode voltage required to produce the dc gate trigger current. Manufacturers sometimes supply curves to indicate V_{gt} and I_{gt} characteristics as a function of temperature.

V_{gfm}, the peak forward gate voltage, is the maximum gate trigger voltage that can be applied without causing damage to the gate–cathode junction.

There are numerous other characteristics. To quote a few, they are average gate power, nontriggering gate voltage, dv/dt rating, and I_{rms}. Explaining these and others is presently avoided because further explanations may degenerate into confusion and loss of fundamental knowledge. With more advanced study and experience the other parameters will then make sense.

8-4 Simple Applications of SCRs

Silicon controlled rectifiers are heavy-current devices. They are employed where controlled amounts of current are required. An example of very high loads is electric welding. Certain electric welds may require perhaps 400 amperes for 12 cycles of 60-Hz supply voltage. This requires precision control in the gate circuit, and indeed it is achieved. On the other hand, an SCR may be used in a low-power circuit such as an electric hand drill. The drill may demand only 1.5 amperes, but the speed required may be perhaps 1/10 of full speed. The SCR is capable of doing just that. Since this latter type of circuit is quite common it bears some attention. Figure 8-7 shows the basic circuit. If we analyze the circuit, one step at a time, the total picture will make sense. The circuit to the left of A and B serves only as a half-wave rectifier whose sole purpose is to provide dc for the gate. Diode D_1 conducts when the ac input makes point A positive with respect to B. The voltage drops are half sine waves and are proportional to the magnitudes of the resistances. The polarities are shown in the figure. C_1 in turn will charge to the magnitude that appears across R_2. When A becomes negative with respect to B (the input sine wave second half) no current will flow through R_1, R_2, and D_1 because D_1 is reverse-biased. C_1 in the meantime starts to

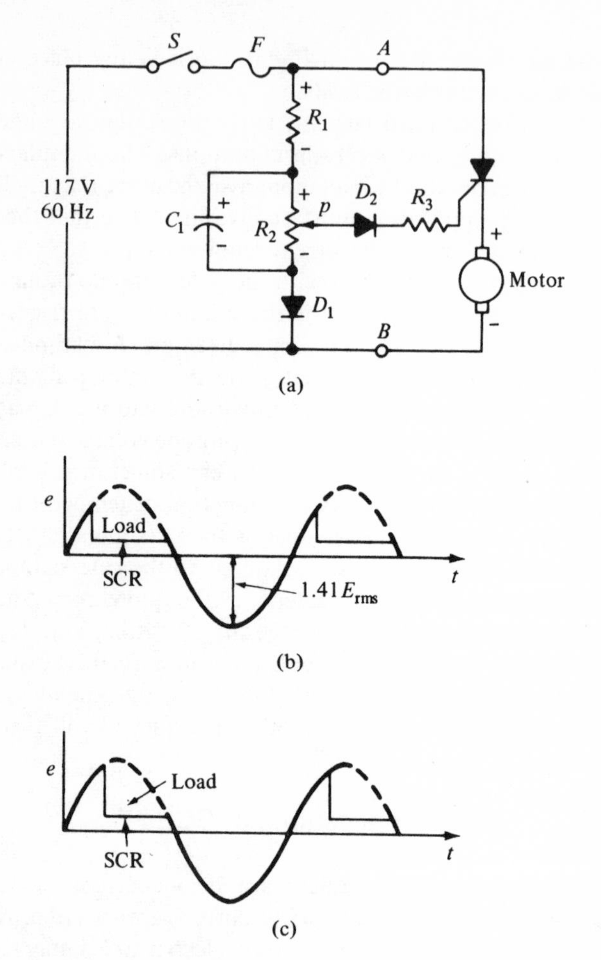

Figure 8–7 (a) SCR motor control circuit, (b) voltage distribution, and (c) voltage distribution with more gate current.

discharge into R_2. This keeps a voltage across R_2, although it is a diminishing one. The C_1 voltage will reduce but at the next half of the input voltage, it will be charged to the maximum it acquired before. Hence, the purpose of the circuit R_1, R_2, C_1, and D_1 is to provide dc voltage for the gate and the magnitude of this voltage depends on the relative values of R_1, R_2, and the source voltage. If a mathematical relationship is desired, we can use proportion to realize that

$$V_{R_2} \simeq V_{\max} \frac{R_2}{R_1 + R_2} \tag{8-2}$$

where $V_{\max} = 1.41\ V_{\text{rms}}$ of the source. If we have 117 volts rms then $V_{\max} = 165$ volts. The purpose of R_1 and R_2 is to produce the desired range of dc voltage across R_2.

If we look at the circuit that remains to the right of A and B, we notice that the SCR has a motor in its anode–cathode circuit. This is the useful load that is to be controlled. Incidentally, the output is dc so a dc motor must be used in the circuit. In the gate circuit, we notice that the gate can be made more positive with respect to the cathode by advancing the control P upward. R_3 is a current-limiting resistor. SCRs and tube equivalents of SCRs are never operated without a load or a limiting resistor in series with the control element. And diode D_2 is in the circuit to prohibit any current flow from anode to gate.

Now for the operation. The motor speeds up whenever more voltage is fed to it. Therefore, advancing R_2 upward increases the dc voltage fed to the gate. Increased voltage means increased current to the gate and, in turn, permits the SCR to fire sooner as the sine wave goes into the positive half. This yields a higher dc voltage to the motor, which causes it to speed up. Once the motor reaches the desired speed the gate–cathode "sees" the motor voltage and the P-control voltage tending to cancel each other out. (If we examine Figure 8-7 and sum up the gate–source voltage, this canceling effect will become evident.) If the speed is sufficiently high, it overides the forward gate bias and the SCR does not fire. This reduces the motor voltage and now the P control takes over to fire the SCR. The net result is that the motor maintains a constant speed because the motor (which is the output) feeds back information to the input, the gate–cathode circuit.

If a reduced speed is required, the P control is turned downward. This sends less voltage to the gate and consequently the ac anode–cathode voltage must rise higher before it fires. Putting it another way, the motor will receive current but for a smaller portion of the sine wave, and thus the voltage is lower. Once this speed is acquired, the voltage across R_2 (from midpoint down) will cancel the voltage across the motor, causing a nonfiring condition. If this results in a slight decrease in motor speed, and therefore less M voltage, the P control once again takes over and firing occurs. Now the system self-regulates at a lower speed.

Figure 8-7(b) shows the voltage that appears across the SCR and load at an early firing or high speed. Figure 8-7(c) shows a later firing or a relatively lower speed. Notice the area of the wave that goes to the motor. The greater the area, the higher the average dc voltage. Notice also that the SCR has a low voltage across it when it is conducting, but it receives the maximum ac voltage (1.41 E_s) when it is not conducting (reverse bias). It is evident that the SCR must be capable of withstanding this peak inverse voltage.

The student may notice that control and current flow occur only at one half of the incoming frequency. It seems rather inefficient not to use the other half of the sine wave. This indeed can be corrected by connecting two SCRs in an antiparallel circuit, as shown in Figure 8-8. Now current can be controlled in either direction, and the current through the load will be alternating. A later chapter will show how another singular device can duplicate these two SCRs.

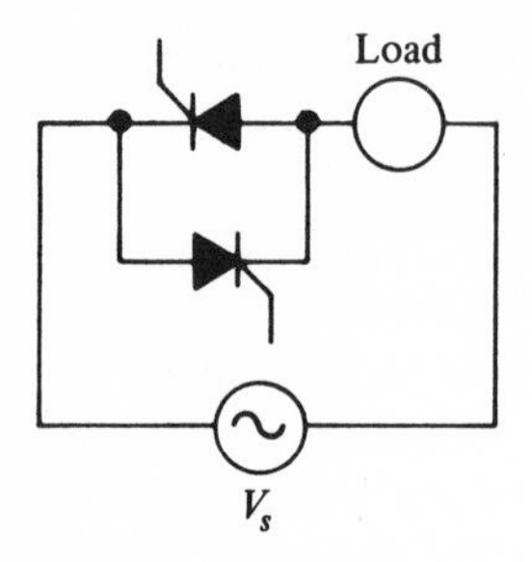

Figure 8-8 SCR connected in an anti-parallel configuration.

Before we conclude this section, it might be worthwhile to integrate some of the devices that have been studied to this point. Let us assume that it is desired to turn lights on whenever darkness sets in. (Utility companies do precisely this.) How can SCRs and UJTs be employed for this purpose? First, and obviously, we must have a photosensitive device that discriminates between darkness and daylight. Second, this device must initiate another circuit that will permit current to flow through a few lamps. This sounds like a job for an SCR. However, a photo device by nature cannot supply sufficient current to trigger the SCR. We therefore must have some kind of intermediate device that can supply the necessary voltage and current. Let us see whether a UJT can fill the bill. Figure 8-9 is an approximate circuit of what we may want. R_2 is a photosensitive resistor whose resistance decreases when light shines on it. When the light intensity is reduced, (dusk) R_2 increases and takes a greater percentage of V_{BB}. This delicate balance of resistors is such that the UJT emitter is now forward-biased and fires. This produces a voltage across R_3 that is fed to the gate of the SCR. The SCR fires and enough voltage appears across the lamps to make them glow. When daylight appears, the resistance R_2 decreases and the UJT cannot be turned on, and therefore the SCR is cut off. Once again we see that bidirectional control would be welcomed. In any case, we managed to coordinate some of the devices that we studied. Although the example we used is elementary, hopefully it will aid in remembering the basic principles discussed.

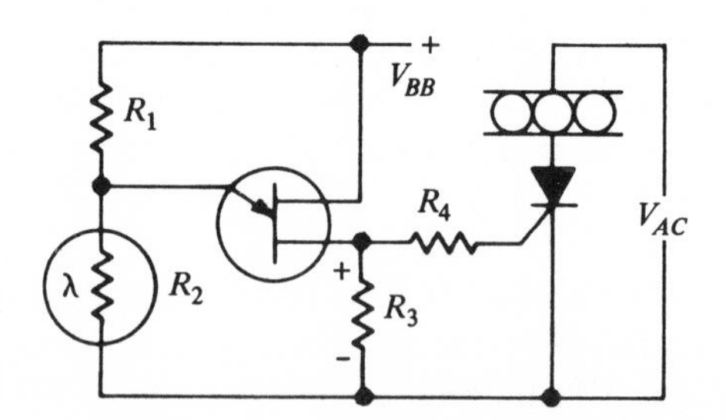

Figure 8-9 Incandescent lamp control circuit using and SCR and a UJT.

8-5 Series and Parallel Operation

Early applications of SCRs required extending their capabilities by connecting them in series or parallel. Placing SCRs in series permitted their use when a single unit did not have the voltage rating. The problem with series connection is that both SCRs may not share the voltage equally, especially if a transient (momentary high voltage) should appear. As a quick example, suppose unit 1 had 500 kΩ in the reverse direction for a particular current and unit 2 had 1.0 MΩ for the same current. If connected in series, unit 2 would "take" proportionately twice the voltage of unit 1 if a high reverse voltage spike appeared. If they shared equally, breakdown might not occur but under the 2-to-1 condition there could be breakdown. Equalization could be achieved if shunt resistors and capacitors were used.

Similarly, the current range can be extended if SCRs are connected in parallel. Once again, unequal sharing of the load current can cause damage. As an example, if one unit "responds" a little faster to certain conditions, it will take most of the current before unit 2 decides to do its share. This short pause may be sufficient to require unit 1 to handle current beyond its capability and therefore burn out. To equalize this condition, resistors are placed in series with each gate.

Presently, series and parallel connections are becoming unnecessary because, as technology advances, voltage- and current-handling capabilities are being extended.

8-6 Summary

In this chapter we found out that a silicon controlled rectifier (SCR) is a four-layer, three-terminal device. It behaves essentially like the UJT, but is capable of handling much higher currents. The anode and cathode are the rectifier terminals, just as in ordinary diodes. The third element, the gate, triggers the rectification process when small but appropriate current enters this element.

The *I-V* characteristic shows that the amount of anode–cathode voltage required to cause conduction decreases as the gate current increases. The peculiar behavior of the SCR is that once it fires the gate loses control. The quickest way to regain control is to reduce the anode–cathode voltage to zero. Instead of using a switch to do that, the interruption is made automatically by the use of an ac voltage between the anode and cathode. This voltage will go to zero twice each cycle and therefore permit a new gate-current adjustment for a larger or smaller anode current. If a relatively high current is desired, the gate current is adjusted so that the SCR fires almost immediately when the anode becomes positive with respect to the cathode, thus yielding a relatively high average current. If the gate current is considerably reduced, the anode–cathode voltage must reach a high value before firing and thereby produce a low average current.

SCRs have made their appearance in the market in small, compact control circuits. Typical present-day use finds them in appliance speed-control and in incandescent light dimming. SCRs have duplicated the performance of equivalent gas tubes and have surpassed them in application because of their large current-handling ability. Also, in recent years, applications have blossomed because the price per unit has been considerably reduced.

Questions and Problems

1. Sketch the symbol of an SCR. Identify the terminals.
2. When a positive voltage is placed from anode to cathode of an SCR, with the anode positive, how many reverse-bias diodes are present?
3. Repeat Problem 2 when the cathode is positive with respect to anode.
4. Sketch a two-transistor equivalent circuit for an SCR. Describe the SCR action with the aid of this equivalent circuit.
5. What primarily determines the amount of anode current flow once an SCR fires?
6. What must be done in the anode circuit so that the gate will regain control?
7. Explain why the use of alternating current between the anode and cathode satisfies the requirements of Problem 6.
8. Sketch two typical forward curves of I_A versus V_{AC} when I_{G2} is greater than I_{G1}.
9. What is meant by holding current?
10. Assume that a sinusoidal voltage appears between the anode and cathode. Assume also that the gate current is such that the SCR fires when the wave passes through its 45° value. Sketch two cycles of input voltage, identify the firing point, and indicate which portion appears across the SCR and which portion appears across the load.
11. What is the purpose of diode D_1 in Figure 8-7?
12. What happens in the D_1 circuit when point A is negative with respect to B in Figure 8-7?
13. In which direction must P be turned in Figure 8-7 so that the motor will increase its speed? Why?
14. Assume that the gate controls in Figure 8-8 are adjusted to fire at a 45° magnitude of a sine wave. Sketch a waveform that shows the voltage across the load and SCRs.
15. What is the peak reverse voltage that the SCR in Figure 8-7 must withstand?

9

The Diac and Triac

9-1 Structure and Basic Operation

There are two devices that are closely related to the silicon controlled rectifier. These are the diac and triac. Understanding them is somewhat simplified if the student is thoroughly familiar with the information in Chapter 8. In brief, a diac is a two-terminal SCR (no gate) and a triac is a bidirectional SCR (that is, once the gate fires the SCR, conduction will occur on both polarities of the input voltage).

A typical unilateral four-layer characteristic is shown in Figure 9-1(a). Notice the similarity between this response and that of an SCR. This characteristic in a sense can be described as a zero-gate-current SCR character-

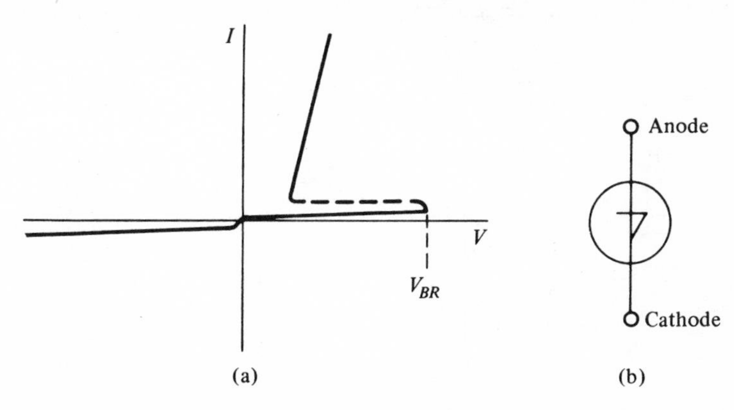

Figure 9–1 (a) Unilateral four-layer characteristic and (b) symbol.

istic and we must rely on the anode-to-cathode voltage to cause breakdown. The only way to obtain other breakdown points is to use another diode, because these diodes have switching voltages from 20 to 200 volts. Therefore, we can conclude that virtually no current flows through the diode until a pulse or a steady voltage greater than the breakdown voltage is applied. In the conduction state the diode has very little resistance. Once the voltage is removed, or reduced to less than the breakdown value, the diode returns to an "open" switch condition or high resistance. From the figure we can conclude that high resistance is achieved in the reverse-bias condition also. Figure 9-1(b) shows the symbol used for this type of switching diode.

A multilayer switching device that has a characteristic similar to Figure 9-2(a) is called a diac. It has the extra advantage of being bidirectional; that is, forward or reverse bias can switch it from high to low resistance, provided the voltage exceeds the breakdown value. These diodes are used primarily in switching circuits, such as in the gate circuits of SCRs. Figure 9-2(b) shows a diac symbol. An application example will follow in a later section.

It was mentioned earlier that triacs are bidirectional SCRs. To understand SCR internal behavior we must refer to Chapter 8. To comprehend triacs we shall simply point out the similarity and deduce its behavior without investigating internal action. First we must refer to the SCR diagram, which is repeated in Figure 9-3(a). Notice that in going from the anode to cathode we "trespass" through *p-n-p-n* regions. This, of course, describes the direction of current through the device once the gate conditions are right. In Figure 9-3(b), as we proceed from terminal 1 to terminal 2, we trespass through *p-n-p-n* regions by selecting the proper starting and stopping regions. This is possible because terminals 1 and 2 touch two differently doped regions. Similarly, if we go from terminal 2 to terminal 1, we pass through *p-n-p-n* regions, as in the SCR. We can conclude from the explanation that a triac's behavior is similar to that of two SCRs connected in parallel, with one SCR's anode and cathode tied to the other SCR's cathode and anode, respectively. Hence, if the gate says "yes" we can expect current to flow in

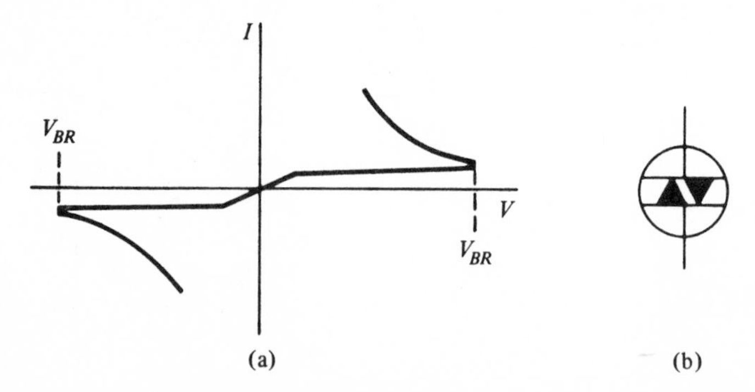

Figure 9–2 (a) Diac characteristic and (b) symbol.

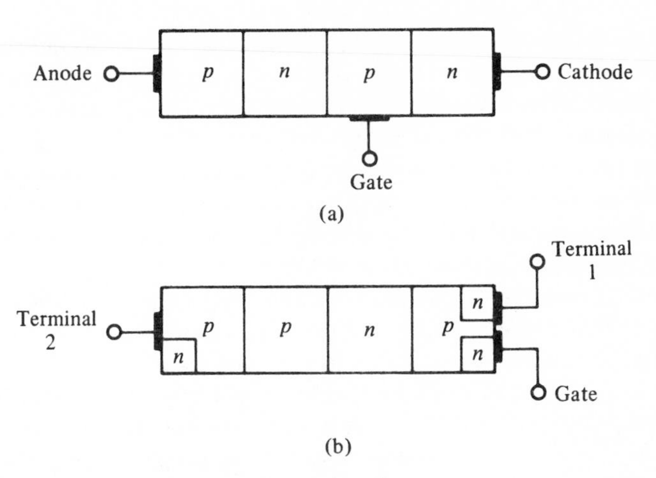

Figure 9–3 (a) SCR diagram and (b) triac diagram.

both directions. Figure 9-4 shows the bidirectional capability of a triac. In addition, the figure shows an SCR characteristic for comparison.

We can also conclude that the terms "forward" and "reverse" are not quite appropriate for a triac. Current can flow in either direction and, since the gate decides the point or voltage at which conduction starts, it is worth our while next to investigate various gate conditions.

9-2 Trigger Modes

In studying SCRs we found out that an SCR anode–cathode current is significantly large once conduction takes place. More correctly, this current depends on the amount of external resistance in the circuit and we should say that the SCR's resistance decreases considerably and therefore has no bearing on the amount of current flow. It essentially is a switch. The gate permits current flow at lower anode–cathode voltages by increasing its

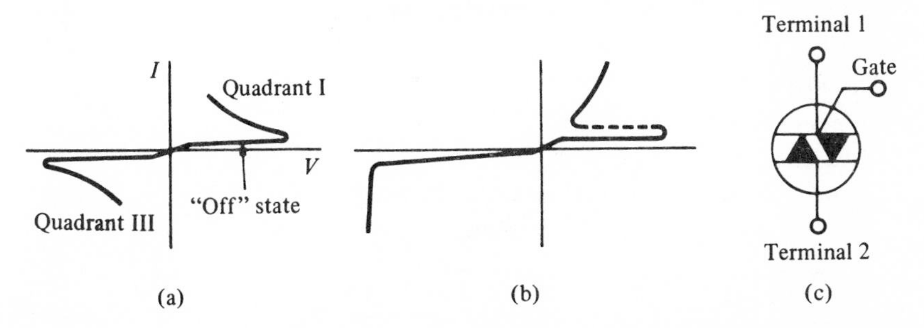

Figure 9–4 (a) Triac characteristic, (b) SCR characteristic, and (c) triac symbol.

current. To do this the gate must be positive with respect to the cathode. If we look at this diagrammatically, we can say that the gate potential (with respect to the cathode) causes a forward-bias condition between the gate–cathode regions. See Figure 9-5(a). Higher biases (more gate current) permitted lower anode–cathode breakdown potentials.

Now let us see what occurs in a triac gate circuit. First of all, there is no cathode. However, the gate voltage is applied between gate and terminal 1. Hence, terminal 1 is the common terminal. If we place a positive gate voltage as shown in Figure 9-5(b), we are essentially duplicating the SCR condition. A forward-bias condition exists in the gate–terminal-1 diode because the *P* material is positive with respect to the *n* material. And if terminal 2 is positive with respect to terminal 1, current will flow as indicated in Figure 9-5(b), or just as in the SCR. This is one mode of operation and we can intuitively expect other combinations. Let us simply reverse the voltage between terminal 2 and terminal 1. Can we expect current to flow through these terminals? Yes, because we still have forward gate–terminal-1 bias and we still trespass through *p-n-p-n* regions as in the SCR. In this case, however, the main current flow is in an opposite direction, as shown in Figure 9-5(c). Since here the main current is opposite to that in the previous case and since the previous case is quite similar to SCR flow we conclude that here the current flow is in the third quadrant, as shown in Figure 9-4(b). The SCR and conditions for the triac in Figure 9-5(b) were operating in the first quadrant.

Let us look at what happens when the gate polarity is reversed and terminal 2 is returned to a positive potential with respect to terminal 1. Investigating Figure 9-5(d) tells us that a forward bias still exists between gate and terminal 1 and gate current will flow as indicated. Hence, main current will flow also through the triac, as shown in Figure 9-5(d). The fourth remaining possibility of operation is shown in Figure 9-5(c). In this case we repeat these conditions except that the voltage between terminal 2 and terminal 1 is reversed. Once again, current will flow through the device because the gate is forward-biased and we traverse through a *p-n-p-n* path. This action is summarized in Table 9-1. From Table 9-1 and the figures we can conclude that we have four triac trigger modes; whenever the current flows from

Table 9-1

Voltage from Gate to Terminal 1	*Voltage from Terminal 2 to Terminal 1*	*Quadrant of Operation*
+	+	I
+	−	III
−	+	I
−	−	III

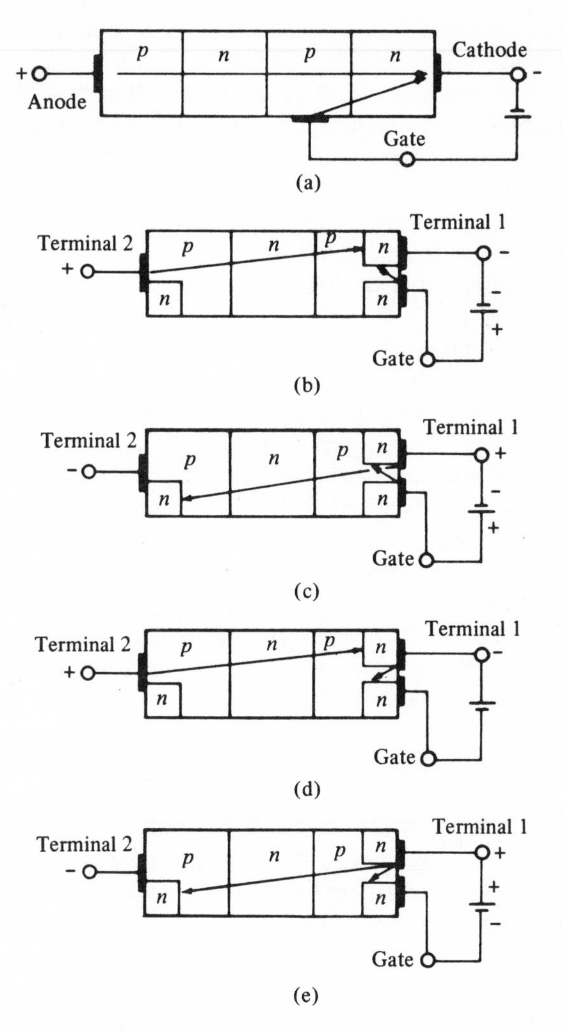

Figure 9–5 (a) Current flow in an SCR triac circuit, with (b) terminal 2 positive and gate *p*-region positive, (c) terminal 2 negative and gate *p*-region positive, (d) terminal 2 positive and gate *n*-region negative, (e) terminal 2 negative and gate *n*-region negative.

terminal 2 to terminal 1 internally we have operation in quadrant I, and when current flows from terminal 1 to terminal 2 the operation is in III. The direction of flow depends on whether the positive side of the battery is connected to terminal 1 or terminal 2. By this we mean that conventional current direction tells us the direction of flow through the triac. Whether the triac fires and permits this conduction depends on the amount of forward-bias current through the gate and the magnitude of voltage between terminals

2 and 1. In any case, it is rather obvious that the triac will be more useful in an ac circuit because both halves of a sine wave will cause conduction and be subject to control. A single SCR is capable of controlling only half a cycle. Indeed, SCRs can do the same job a triac can do, but this requires the connection of two SCRs in an inverse parallel arrangement—a slightly more complicated circuit.

9-3 Triggering Considerations

We recall that once a thyristor (SCR or triac) fires, the gate loses control until either the anode–cathode or terminal-2–terminal-1 voltage is reduced to zero volts. Therefore, it seems logical that triggering can be effected by a pulse of voltage versus a steady quantity. After all, once the SCR or triac fires who cares whether a voltage still remains on the gate? Therefore, the question is asked, how big and how long should this pulse appear between the gate and terminal 1? To answer this question we must investigate thyristors a little further. First of all, current flows through the gate circuit to turn the thyristor on. If this is so, power is consumed in the device. Indeed, there is a manufacturer's maximum that cannot be exceeded. We all are aware that calculating power requires finding the product of volts and amperes. This is quite simple when a steady-state condition exists. However, if V and the resulting I are pulsed, how do we know whether or not the average power rating for the gate is exceeded? (It has been mentioned earlier that most electronic devices have maximum power ratings, but that these ratings can be exceeded for short durations.) In addition, if current pulses are used, how long should the pulses be to guarantee that the thyristor will turn on? To help in selecting the right pulse duration and not exceeding the power, manufacturers usually supply gate characteristic curves. Figure 9-6(a) shows the relationship between turn-on time and gate current in a

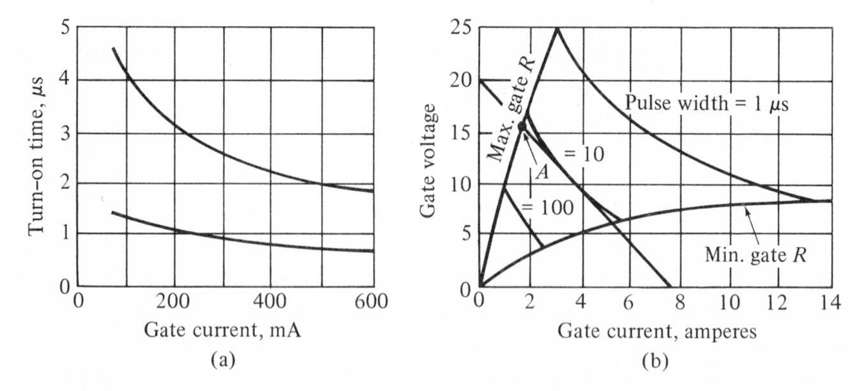

Figure 9–6 (a) SCR turn-on time characteristic and (b) gate pulse-duration characteristic.

thyristor. Two curves are shown because thyristors have a spread in this response. Notice a low current has a long turn-on time and a high current triggers the device faster. Usually a higher gate current is used to guarantee turn-on, particularly if thyristor units are changed. Once the gate current is known, we use Figure 9-6(b), which shows the relationship of gate voltage to gate current or, in other words, the resistance of the gate. Once again, two curves are shown for gate resistance because thyristors of the same family have a range of values. Notice that pulse widths are also indicated. To obtain the required pulse width, we simply go vertically with the required gate current, obtained from Figure 9-6(a), and stop at the intersection of this gate current and the maximum gate resistance. The pulse duration must be less than the nearest estimated pulse-width line. Though somewhat crude, this practice is acceptable.

In all applications a resistor is placed in series with the gate. However, we must guarantee that the estimated gate current flows through the gate to fire the thyristor. An external resistor will, indeed, lower the current. To compensate, the pulse source voltage is selected so that source voltage divided by gate plus external resistance is equal to the firing current. Putting it another way, a line is drawn through this intersecting point (gate current–maximum gate resistance) that intercepts the ordinate. See point *A* in Figure 9-6(b). Obviously, many lines can be drawn through this point, but the voltage required is the ordinate's intercept and the series resistance for this voltage is the slope of this line. A different ordinate intercept yields a different slope, which means a different series gate resistance. But no matter what combination is used, they all go through the required gate current. For the line drawn in Figure 9-6(b), the pulse source voltage should be 20 volts, the maximum width should be approximately 10 μs and the series resistance is 20/7.5 or 2.6 ohms. Notice that the pulse width is quite large in comparison to the quoted turn-on times in Figure 9-6(a)—in order, of course, to guarantee firing.

9-4 AC Switching

In Chapter 8 a circuit that triggered an SCR was studied. In this circuit, control of firing occurred on one half of the sinusoidal input. It is conceivable that it might be desirable to control both halves despite the rectifying capability of the SCR; that is, we might wish to control an ac load. A typical ac load application is the dimming of incandescent lamps. Figure 9-7(a) shows how two SCRs would be needed. For simplicity, the trigger circuit is not shown, and notice also the waveshape at the load. Changes in the trigger circuit produce changes in the output wave.

Reviewing SCR principles we can say that a high gate current will cause the SCR to fire sooner. An earlier firing means that more of the sine wave will appear across the load, and if the load is a bulb, it will glow brighter. It

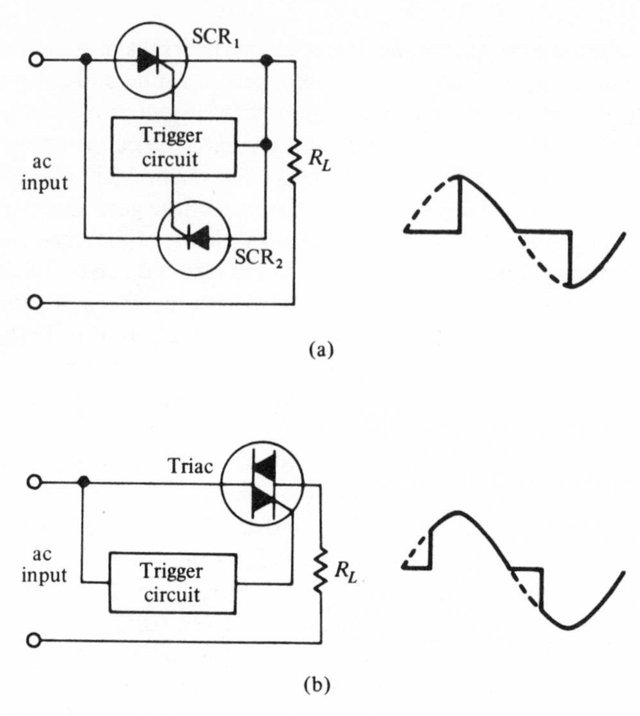

Figure 9–7 Variable ac control using (a) two SCRs and (b) one triac.

is worth repeating that an SCR acts like a switch, and whenever a switch is open in a simple series circuit, all of the input voltage appears across the open terminals. When the switch is closed, current flows; no voltage appears across the switch but instead all of the input voltage appears across the other devices in the circuit, in this case the load resistance. The extra benefit gained in an SCR switch is that we can control how much of the input appears at the output. Now what does appear across the load, ac or dc, depends on what is in the circuit. In Figure 9-7(a), we expect ac because two SCRs are connected in an inverse parallel arrangement. Figure 9-7(b), on the other hand, shows how a single triac is connected to do the same job. This is possible because the triac can be turned on by a positive or negative voltage, and the main current can flow in either direction.

In both of the above circuits the triggering circuit is shown in block form. Perhaps now would be an appropriate time to investigate these to see how ac can be used to turn on thyristors, in light of the fact that ac by nature changes its amplitude with time. Remember that in most applications it is beneficial to keep components to a minimum and if variable ac is required, then a simple control is quite appropriate.

The simplest method of firing thyristors with ac is shown in Figure 9-8. The same voltage that appears across the heavy-current terminals of the thyristor also appears between the gate and cathode of the SCR and between

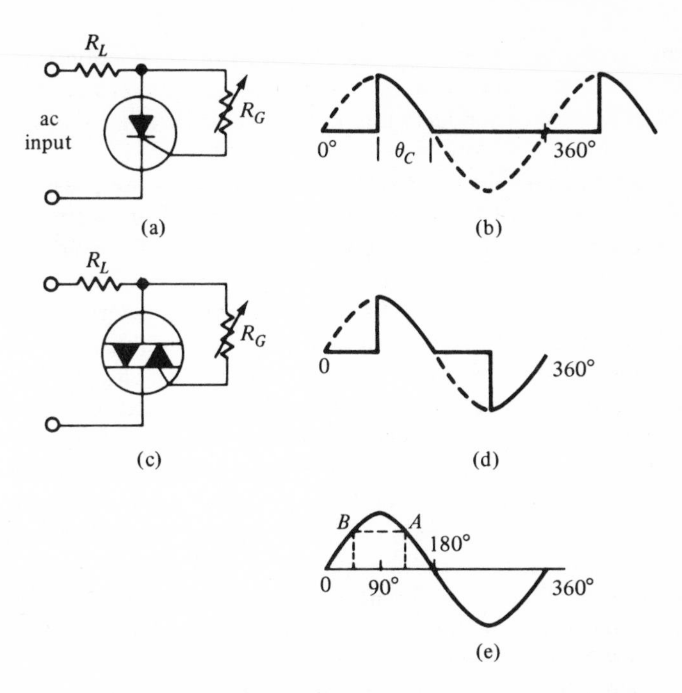

Figure 9–8 (a) Simple SCR gate control and (b) resulting voltage across load. (c) Triac gate control and (d) resulting voltage across load. (e) Location of similar firing voltages.

the gate and terminal 1 of the triac. When R_G is low, conduction starts early and when R_G is large the input voltage must go relatively high before conduction occurs. Figures 9-8(b) and 9-8(d) show the voltage across the load. These two figures show the smallest load voltage possible for this type of firing. By this we mean that later firings are not possible. To further clarify this concept, Figure 9-8(e) is used. Suppose we wanted to delay firing until point A of the sine wave. Indeed, a smaller current would result because conduction would only last from A to 180°. A small sliver of voltage would occur across the load. However, a thyristor fires at the first voltage it sees that is equal to A. If a sine wave proceeds from zero the thyristor will see this magnitude at B and, hence, will fire. Therefore, we can conclude that since voltages from 0 to 90° are duplicated in the 90 to 180° range, no firings can be delayed beyond 90°. Similarly, with triacs no firings can occur from 90 to 180° and from 270 to 360° of a sine wave. Since conduction angle is preferably used over firing angle we can define θ_C (conduction angle) in terms of θ_F (firing angle). For half-wave (single-SCR) circuits,

$$\theta_C = 180° - \theta_F \tag{9-1}$$

For full-wave (triac) circuits,

$$\theta_C = 2(180° - \theta_F) \tag{9-2}$$

We can use the equations to check the limits of conduction for a single SCR and a triac. Since the SCR can be fired as early as zero degrees and as late as 90° the range of conduction is 180° down to 90°. For a triac it is 360 to 180°.

In some applications the fact that firing beyond 90° is not possible may be a drawback because smaller values of dc (single SCR) or ac (triac) may be desired. The incandescent lamp dimmer is an example. To achieve this end, the simplest way is to use a resistance–capacitance network, commonly called a phase-shift circuit. Figure 9-9 shows how this network circuit is used in a triac control circuit, which could be feeding a variable-speed motor or a dimming incandescent lamp. First of all, notice the *RC* circuit. It is directly across the source. The voltage that appears across these two components depends on the position of *R*. A large *R* means that the voltage across *C* is small, and vice versa. This of course is true on a steady-state (rms values) basis and on an instantaneous basis. Second, the voltage that appears across the capacitor is the voltage that is fed to the trigger diode, which is in series with the triac gate. Therefore, the capacitor voltage rises as the input rises and when the magnitude is sufficient, the diode conducts, and the capacitor discharges and triggers the triac. Conduction will occur until the input voltage goes to zero. (We must remember that the gate loses control once the triac fires and restoration of control occurs when the voltage between terminals 1 and 2 reaches zero.) When the input voltage reverses direction, the same thing happens because the trigger diode is bidirectional and the triac fires with either polarity.

The foregoing explanation may seem somewhat easy to comprehend. However, it has not been mentioned how later firing angles (shorter conduction angles) can be achieved. After all, this is the main reason for going into an *RC* circuit. Without getting too involved we can explain this feature with some phase-relationship sketches. The thing to remember is that there are two voltages of importance: one is the main input voltage, which feeds the load; the other is the capacitor voltage, which feeds the triac gate.

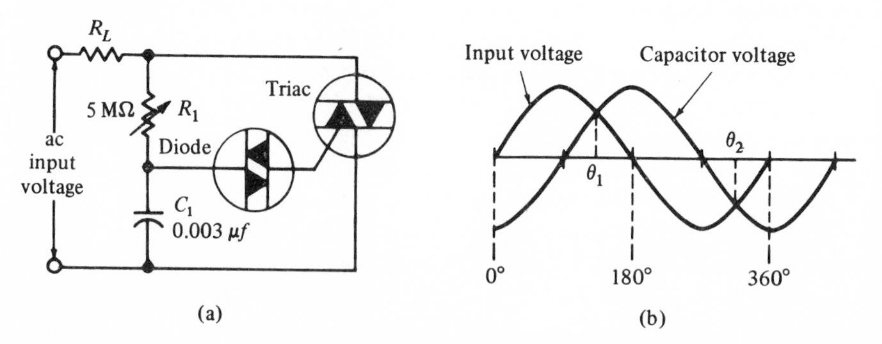

Figure 9–9 (a) Variable phase-shift control to triac and (b) phase-shift voltages.

Figure 9-9(b) shows how they are related to each other. Notice that the capacitor voltage lags the line voltage. When the magnitude of the capacitor voltage reaches the firing point, the main terminals conduct. This of course may not occur until the main voltage has passed the 90° value. See θ_1 in the figure. A smaller resistance, R_1, will decrease the phase lag angle and the firing will occur sooner. This means larger values of ac will pass through the load. Incidentally, θ_1 is a firing angle; the conduction angle is the distance between θ_1 and the 180° (zero-volts) point for the line voltage.

The above circuit does not perform exactly as explained in the previous paragraphs. A problem that exists is that once the capacitor discharges it does not follow in a sinusoidal fashion. This does not allow the triac to fire at small conduction angles. Suffice it to say that manufacturers have designed double *RC* circuits to minimize this problem. This means that another *RC* circuit appears across C_1 and the second capacitor voltage is fed into the diac–triac circuit. This way, more stability is added.

9-5 Heat Sinks

Most electronic devices generate heat. A simple illustration is the television set. The removal of heat is always a problem and manufacturers always take this into consideration. In the case of a TV set components are spaced, and ventilation is designed in, so that the components do not exceed their temperature limits.

The source of the problem is easy to understand. Any time a device has some equivalent resistance, power will be dissipated when current flows through it. This includes tubes as well as transistors. And how much power a device can take without burning out depends primarily on physical size. This, of course, forebodes low power capability for semiconductor devices for indeed they are small. However, one way out of the dilemma is to arrange for heat to be conducted away from a small semiconductor pellet. This is precisely what is done. A collector in a given transistor may be attached to the transistor metal case and the case in turn may be in contact with a metal chassis. The case and chassis will indeed conduct heat away from the tiny crystal. Sometimes conditions are such that a piece of metal is used instead of the chassis. This chunk of metal is called a heat sink. The three most common heat-sink materials are copper, aluminum, and magnesium. Each has its advantages; but since cost is generally the dictating factor, aluminum is most frequently used.

Heat sinks have different coatings. Some assist in dissipating heat, others act as electrical insulators, and still others protect against corrosion. Each depends on where the transistor or semiconductor device is ultimately going to be used. As an example, untreated aluminum can radiate heat well enough; however, if corrosion is a problem then the aluminum requires anodizing. In any case, heat sinks are made commercially and can be purchased provided the correct specifications are given. Heat sinks come in various shapes and forms. Figure 9-10 shows a sketch of a typical high power heat sink. Notice the fins. These fins dissipate a large amount of heat because the greater the surface area, the greater the heat-dissipation ability.

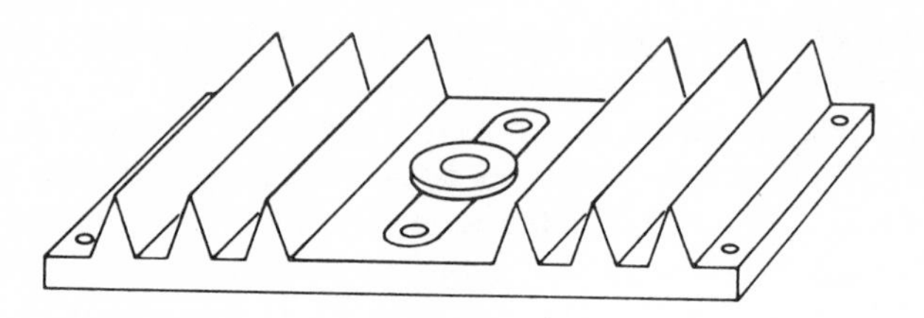

Figure 9–10 Commercial heat sink sketch.

Heat sinks are used with semiconductor devices that are subject to relatively high power and where natural cooling is not adequate. These devices are: Zener diodes, conventional diodes, thyristors, and transistors. In other words, any device that develops power may require some heat-sinking. Most of the manufacturers supply derating curves for their power devices. Basically, these curves tell you what power should not be exceeded if the ambient temperature is greater than room temperature or 25°C. Figure 9-11 shows a typical derating curve. The 1.0-watt rating is the quoted value in the manual. If the installation subjects the device to an ambient temperature of 75°C, then it should not be designed to dissipate more than approximately 0.66 watt. Otherwise, burnout may result.

As far as heat sinks are concerned, determining the proper area is a mathematical science. On the other hand, some manufacturers supply nomo-

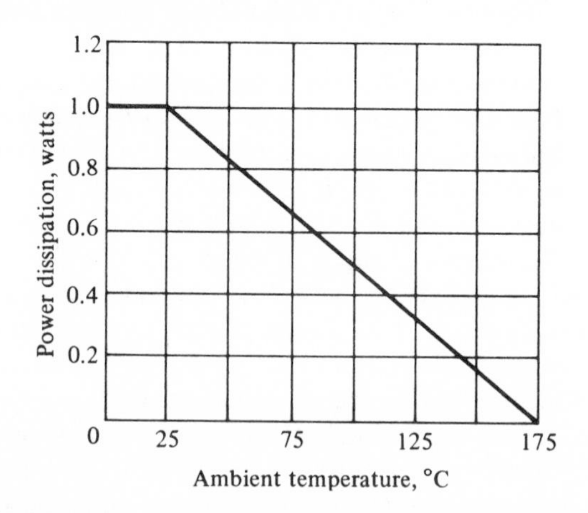

Figure 9–11 Semiconductor derating curve.

graphs which make the selection of a heat sink a little easier. In either case, it is important to have some perception of the underlying principles. Between the pellet (semiconductor) and the atmosphere are elements that make thermal contact with the high-current terminals. In the case of a transistor, it is the collector. The collector, in turn, is tied to the case; the case, in turn, is in contact with perhaps an insulating washer and then the heat sink. So between the pellet and the heat sink the heat is conducted and then radiated to the atmosphere.

Each of these elements does its share of dissipating the heat; collectively, the elements must not permit the semiconductor to exceed its maximum temperature, which is specified by the manufacturer. How much each element contributes depends on a factor called thermal resistance, which, by definition, is the temperature difference between two points when 1.0 watt is dissipated. In any case, these thermal resistances are expressed in degrees centigrade per watt, and, for a given installation, are added together, as series resistances are added. Mathematically, the total thermal resistance is

$$\theta_{ja} = \theta_{jc} + \theta_{cs} + \theta_{sa} \tag{9-3}$$

where

θ_{ja} = total resistance, junction to ambient
θ_{jc} = thermal resistance, junction to case
θ_{cs} = insulator thermal resistance, case to sink
θ_{sa} = heat sink thermal resistance, sink to ambient

In most cases, θ_{jc} and θ_{cs} are given and it is the thermal resistance of the heat sink that is required. This resistance is then related to the physical size or area of the heat sink.

However, it is important to correlate this total thermal resistance to the power dissipation ability of a semiconductor device. Notice that if we multiply θ_{ja} by a power quantity (in our case, the power rating of a transistor) the result is temperature. Therefore, the temperature differential that can exist between the semiconductor and the ambient temperature can be expressed as

$$\theta_{ja}P_s = T_{j(\max)} - T_a \tag{9-4}$$

where P_s is the power handling ability of the semiconductor and $T_{j(\max)}$ is the maximum temperature the semiconductor can withstand. This latter quantity is given by the manufacturer. Thus the approach to heat sink determination requires the use of Equation (9-4), where θ_{ja} is evaluated. Then Equation (9-3) is used to determine θ_{sa}. This quantity is related to the physical size of a heat sink.

Although heat sinks are commercially available, any mass of conducting material can be used provided the proper thermal resistance is achieved.

9-6 Summary

In this chapter we studied the behavior of diacs and triacs. The diac, we noted, behaved just like an SCR that had zero gate current; that is, the diac conducted after a certain voltage across it was developed. In addition, the device was bidirectional; either a positive or negative voltage fired this two-lead device.

The triac, that followed, seemed a natural development of SCRs. This device is a bidirectional SCR. The amount of ac flow through the device depends on the gate current. More gate current permits the triac firing earlier in a sinusoidal input. Earlier firing of both halves of a sine wave yields higher outputs.

Following the study of triacs, we looked into the method of firing thyristors. In one case we revealed that pulses to the gate can turn on a thyristor. However, these pulses must have sufficient duration and, at the same time, must not exceed the power rating of the gate. On the other hand, we investigated a means for extending the firing range when ac is used in a triac. Essentially, a phase-shift circuit is used in the gate circuit so that firing may occur beyond the 90° point of the incoming power sinusoid.

Finally, we investigated the underlying principles of heat sinks. Basically, the semiconductor is kept below burnout temperatures by having the heat conducted away to the atmosphere. Each thermally conductive element contributes to this conduction. The only variable element in this circuit is the heat sink, the area of which is usually calculated or determined with the aid of a nomograph.

Questions and Problems

1. Sketch an *I-V* curve for (a) an SCR and (b) a triac.
2. What are the four layers in an SCR as one proceeds from anode to cathode?
3. Does conduction occur between anode and cathode in an SCR when the gate is forward- or reverse-biased with respect to cathode?
4. Trace the alternating current from terminal 2 to terminal 1 (through four layers) when the gate is (a) positive with respect to terminal 1 and (b) negative with respect to terminal 1.
5. What two important factors must be considered for a gate to be set into the firing condition?
6. (True or false.) In triggering a triac with a gate pulse a low current has a long turn-on time and a high current a short turn-on time.
7. When ac appears across an SCR firing can occur from ____ degrees to ____ degrees. Firing cannot be delayed beyond ____ degrees.
8. In the case of a triac, firing cannot occur from ____ to ____ degrees and ____ to ____ degrees of the input wave.

9. To delay firing beyond 90 degrees a ____ ____ circuit is used, which is an *RC* circuit that causes the gate voltage to (lead) or (lag) the main input voltage.
10. Why are heat sinks used?
11. If a heat sink is not used in a low-power transistor, how is the heat dissipated?
12. Using Figure 9-11, what design dissipation is permissible when the ambient temperature is (a) 25°C? (b) 80°C?
13. What is thermal resistance?
14. How is thermal resistance related to the temperature of a semiconductor?

10

Light-Operated Devices

10-1 The Nature of Light

It is common knowledge that radio and television are possible because electromagnetic energy is transmitted through space. This electromagnetic phenomenon is associated with high-frequency alternating current; a typical frequency for AM radio transmission is 10^6 hertz. Light, on the other hand, also is energy and moves through space, but at a frequency of about 10^{15} hertz. The reason we call this energy "light" is that our eyes respond to these frequencies. If the same kind of energy passes through space but at a slightly lower frequency than light frequencies, we call it heat. Radiant heat from the sun falls into this category. So we see that heat and light energies are similar in nature to radio frequencies because of their radiation qualities and the frequency phenomenon.

To get an overall picture of the range of radiant frequencies, a chart or frequency spectrum is displayed in Figure 10-1. These limits of each region are approximate because the chart is purposely simplified. It is worth mentioning other notable frequency ranges because these are frequently heard.

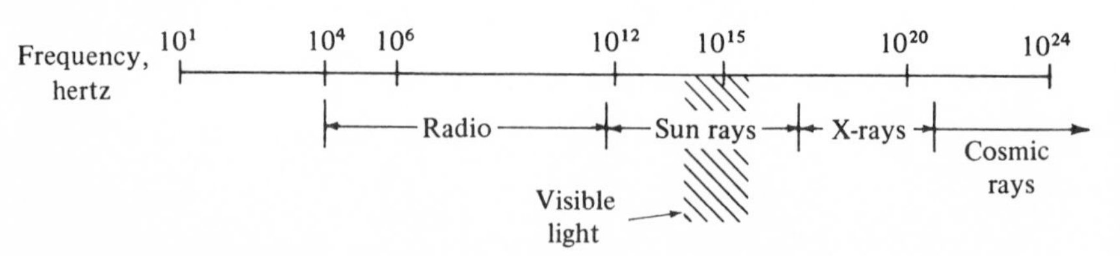

Figure 10–1 Frequency spectrum of electromagnetic energy.

One example is infrared. This falls immediately to the left of the light range in Figure 10-1. Another one is ultraviolet, which falls immediately to the right of the light region. Gamma rays straddle the X-ray and cosmic-ray regions. In any case, all these electromagnetic radiations behave similarly and indeed travel through space at the same speed. This speed is about 186,000 miles per second or, in the metric system, 300,000,000 meters per second.

The speed of these waves leads to a mathematical relationship that says

$$\lambda = \frac{V}{f} \tag{10-1}$$

where λ is the wavelength, V is the velocity in meters per second (3×10^8), and f is the frequency in cycles per second (or hertz).

In this definition, units for λ were purposely omitted because we hope to extract this information by substitution and perhaps make it more meaningful. Hence, if we substitute units in Equation (10-1)

$$\begin{aligned} \lambda &= \frac{V}{f} \\ &= \text{meters/second} \div \text{cycles/second} \\ &= \text{meters/second} \times \text{seconds/cycle} \\ &= \text{meters/cycle} \end{aligned}$$

Now the question arises, what does λ represent when its units are meters/cycle? The units imply that the wave (heat, light, or radio) has traveled so many meters for each cycle. Putting it another way, the distance between each wave crest in space is so many meters. Consider the following example:

EXAMPLE 1

What is the distance between crests of a radio wave whose frequency is 3×10^8 cycles per second?

SOLUTION

$$\begin{aligned} \lambda &= \frac{V}{f} \\ &= \frac{3 \times 10^8 \text{ m/s}}{3 \times 10^8 \text{ cycle/s}} \\ &= 1 \text{ m/cycle} \\ &\simeq 40 \text{ in./cycle} \end{aligned}$$

EXAMPLE 2

Repeat Example 1 for a radio wave that is oscillating at 100,000 hertz.

SOLUTION

$$\lambda = \frac{V}{f}$$
$$= \frac{3 \times 10^8}{1 \times 10^5}$$
$$= 3 \times 10^3 \text{ m/s}$$
$$\simeq 9850 \text{ ft/s}$$

Notice the difference in answers; in one case we have approximately 3 feet and in the other nearly 10,000 feet. It is this number, λ, that determines the size of a radio-frequency antenna. This again is a specialized area of study which is considered outside the scope of this book.

From the question it is obvious that as the frequency increases, λ decreases. It becomes quite small when we reach light frequencies. As an example, if we reach 3×10^{15} cycles per second, the answer for λ is 1×10^{-7} meters/cycle. In any case, when we wish to address ourselves to a particular frequency of light we can use the frequency (3×10^{15} Hz in this case) or the corresponding wavelength (1×10^{-7} meters). This latter approach is precisely what is done. When light waves are discussed, scientists prefer to use the wavelength and not the corresponding frequency.

Now that we know something about the general behavior of waves, let us zoom in on the region that we call light. First of all, each color of light has a different wavelength, or frequency. The light region in Figure 10-1 goes through the colors red, orange, yellow, green, blue, and violet as the frequency increases (or the wavelength decreases). One can readily see why the two regions outside the visible light region are called infrared and ultraviolet. Both border similar colors in the visible range.

Measurements have shown that the wavelengths human eyes respond to ranges from 0.38×10^{-6} to 0.76×10^{-6} meters. Physicists and engineers usually prefer to drop the 10^{-6} factor and introduce the unit called a micrometer (or micron); that is,

$$10^{-6} \text{ meter} = 1 \text{ micrometer}$$

Hence, we can say the human eye responds to 0.38 to 0.76 μm. We must also be aware that human eyes do not treat each color equally. Eyes are more sensitive to yellow and yellow-green colors than to blue or red. Figure 10-2 adequately displays relative response or this sensitivity versus equally intense wavelength or color of light. Some of the colors are placed in the figure. Notice also that these colors are identified in an ascending order of wavelength. The colors would be reversed if the curve was arranged in a frequency-ascending order, as in Figure 10-1.

Some texts identify wavelength in angstroms. The relationship is

$$1 \text{ Å} = 10^{-10} \text{ meters}$$

Therefore, the range of eye response is 3800 to 7600 Å.

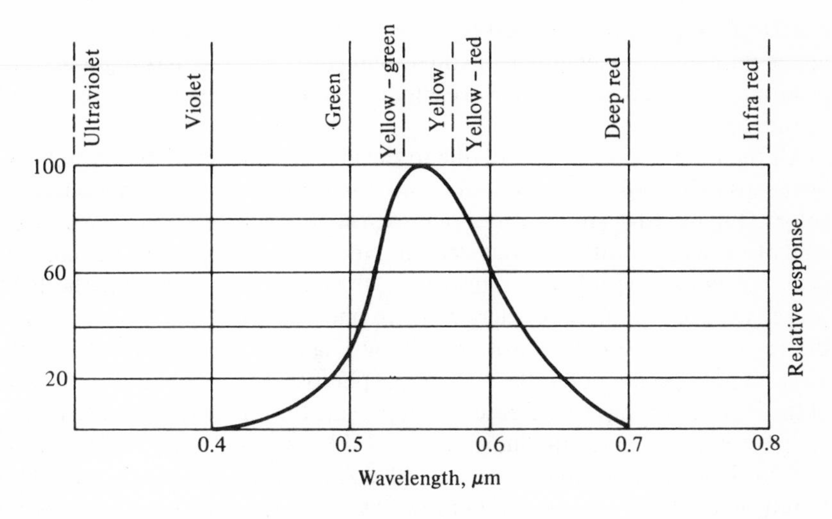

Figure 10–2 Relative response for the human eye.

There are many sources of light; examples are the sun, fluorescent lights, mercury lights, and incandescent lights, and of course burning substances such as candles or kerosene. Each has its own characteristic shading or type of illumination. Some tend to be bluish in color, like many fluorescent lights; others tend to be yellowish, such as a burning candle. In any case, the color of light depends on the wavelengths that are present. In the case of a fluorescent light, wavelengths near the lower end of Figure 10-2 would predominate and all other would be diminished or not be present. In the case of a burning candle, a narrow range of wavelengths near the yellow region would be present. Figure 10-3 shows a spectral distribution for a tungsten lamp. This figure shows that a big spread of wavelengths is present, but only

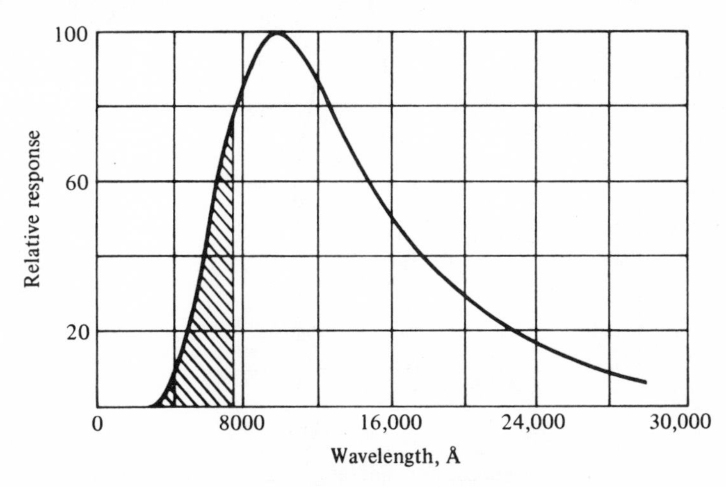

Figure 10–3 Radiant energy distribution of a tungsten lamp.

a narrow range (3800 to 7600 Å) is perceived by the human eye and the upper range of wavelengths is favored. Sunlight, however, has a good composition of all eye-sensitive wavelengths. Sunlight, incidentally, is called "white light."

All these shades of colors are due to the characteristic material. Similarly, light-sensitive solid-state materials tend to respond to their characteristic colors. Figure 10-4 shows three photosensitive solid-state materials. Each tends to peak at a different wavelength. Notice that silicon and germanium peak beyond the visible eye range. It is, therefore, necessary that the light source used to stimulate or energize these photodevices have wavelengths in the region of maximum sensitivity. If by chance a light source has only wavelengths that are not at the peak of the photosensitive device, the intensity of the light source has to increase to compensate for the loss of this sensitivity. As an example, Figure 10-4 indicates that CdS has a maximum response near 6000 Å. However, if a light source of 5000 Å is present, its relative response is only 40 percent. This means that the 5000 Å source must have an intensity $2\frac{1}{2}$ times greater than that of the 6000 Å source just to have an equal response. In any case, perfect source-to-receiver matching is nearly impossible. However, the receiver (the photosensitive device) can still function, provided that a characteristic wavelength is present with sufficient intensity.

10-2 Photoconductive Cells

Three general classifications of photodevices are: photoconductive, photovoltaic, and photoemissive. The effect of light on photoconductive cells is to increase conductivity. This means that as light intensity increases (within the proper wavelength range), the resistance of the photocell de-

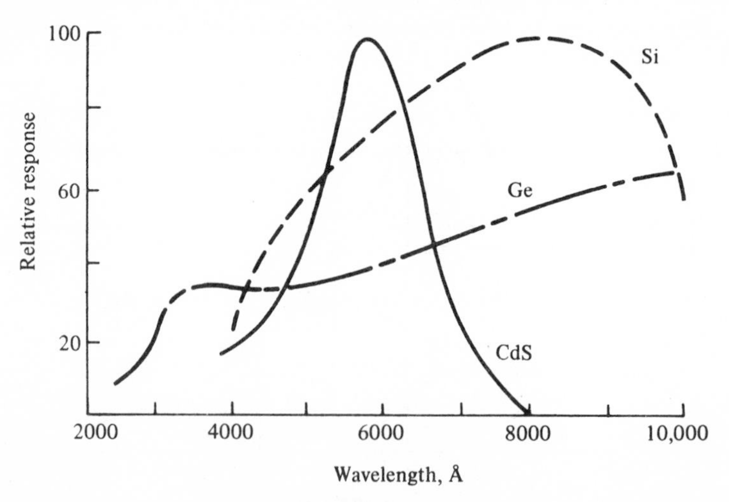

Figure 10–4 Relative response for different light-sensitive materials.

creases. These cells are also referred to as photoresistive types. A resistance ratio of 100/1 to 10,000/1 is possible from a dark-to-light condition. Devices that produce this effect are made from cadmium sulfide, cadmium selenide, selenium, germanium, and silicon.

Figure 10-5(a) shows the basic construction of a photoconductive cell. Cell resistance is determined by electrode geometry. Essentially, the gap between electrodes is decreased for lower resistance. However, closer electrodes limit the voltage capability because breakdown is a function of electrode spacing. This, of course, is an important practical consideration, in that closely spaced electrodes can be used only in relatively low-voltage circuits. Manufacturers also make photoconductive cells by depositing a thin layer of semiconductor material on a substrate and then plating metal electrodes in a comblike fashion, as shown in Figure 10-5(b). This technique permits lower resistances and efficient use of element area.

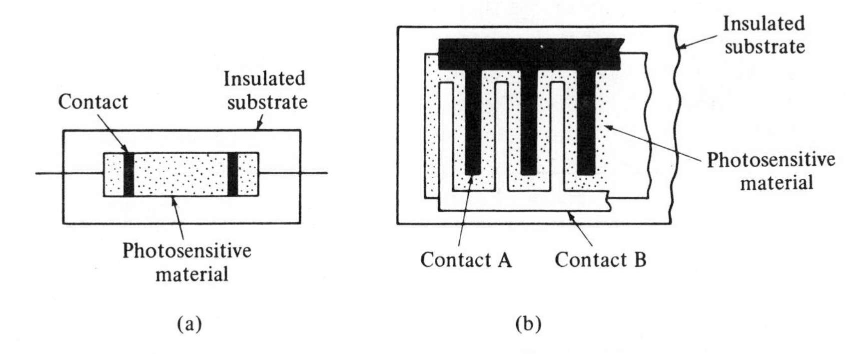

Figure 10–5 (a) Simple photoconductive cell. (b) Long-pathed photoconductive cell.

Manufacturers usually supply spectral response curves for their photoconductive cells. The shape of the curve may vary for products of different manufacturers even when the same type of material is used. This is so because the response is dependent on dopants, impurities, and particle size, and each manufacturer has his own control over these variables.

A simple circuit that demonstrates light sensitivity of a photoconductive cell is shown in Figure 10-6. Here a physical phenomenon, the amount of

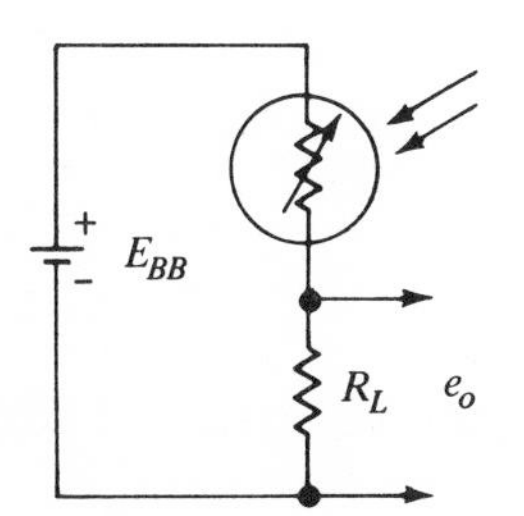

Figure 10–6 Simple circuit employing a photoconductive device.

light, is converted to a corresponding change in voltage across R_L. The greater the intensity of light, the smaller the photoresistance and thus a larger portion of E_{BB} is present across R_L. This fundamental circuit is shown because it will help to show the difference between a photoconductive and a photovoltaic cell.

A more interesting circuit that also incorporates a unijunction transistor is shown in Figure 10-7. Here the photoresistor and capacitor form an *RC* circuit. The action is similar to the one describing a relaxation oscillator. However, the frequency is determined by the intensity of light. The greater the intensity, the smaller the *RC* time constant; hence capacitor C_1 will build up faster to a firing potential for the UJT. The output is a series of pulses that increase in frequency as the light intensity increases, and vice versa. Feeding this information to a digital counter would produce a means of correlating intensity with a numeric readout.

It was previously mentioned that the most important characteristic is the spectral response. Other ratings are resistance ratio, peak voltage, maximum current, maximum power dissipation, and response time. The last parameter is a time value that indicates how quickly a device responds to light. These time constants may range from a few microseconds to 30 milliseconds. Each photosensitive material has its own characteristic time response; application determines whether a given photoconductive cell with its time constant is adequate.

10-3 Photovoltaic Cells

Photovoltaic cells are devices that develop an electromotive force when illuminated. The most common is the solar cell. The oldest is the selenium cell, which has been on the market for approximately 40 years. Because of its simplicity it has been used in portable light meters, in which the cell is connected directly across a meter. When an object reflects light,

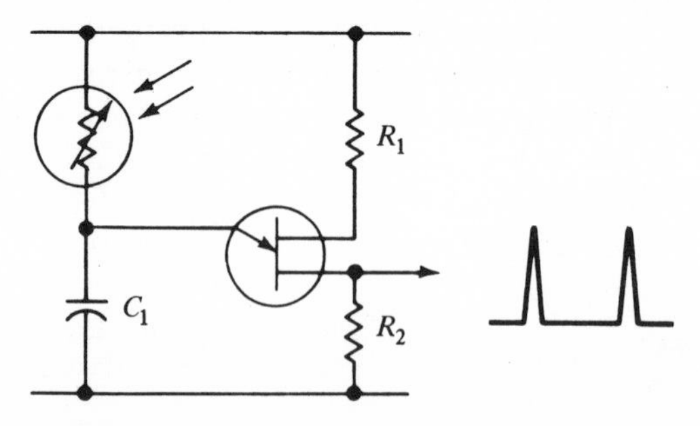

Figure 10–7 Simple light circuit employing photoconductive cell and UJT.

the meter deflection rises. The meter of course is calibrated in units that assist the photographer. The open-circuit voltage varies logarithmically with illumination to a maximum of approximately half a volt. A selenium cell is merely a thin layer of selenium between a metal plate and a transparent electrode.

A silicon solar cell is used primarily to generate power in space satellites because the conversion efficiency of silicon cells is greater than that of selenium. Efficiency is the ratio of output electric power to input solar power. Efficiencies of 10 percent are easily obtainable at present. The basic solar-cell construction is shown in Figure 10-8(a). This rectangular type lends itself to stacking, an arrangement in which the total voltage equals the sum of that of the individual cells. To increase the current capability, cells are connected in parallel. This, of course, is how more power is obtained from ordinary power sources, such as dry cells. Figure 10-8(b) shows a typical *I-V* characteristic. Condition 2 in the figure represents an increase in light intensity over condition 1. I_{SC} represents the short circuit current. This region is relatively linear and the silicon solar cell can be used as a light meter (just like the selenium cell) if the meter has relatively low resistance. V_{OC} is the open circuit voltage ($I = 0$).

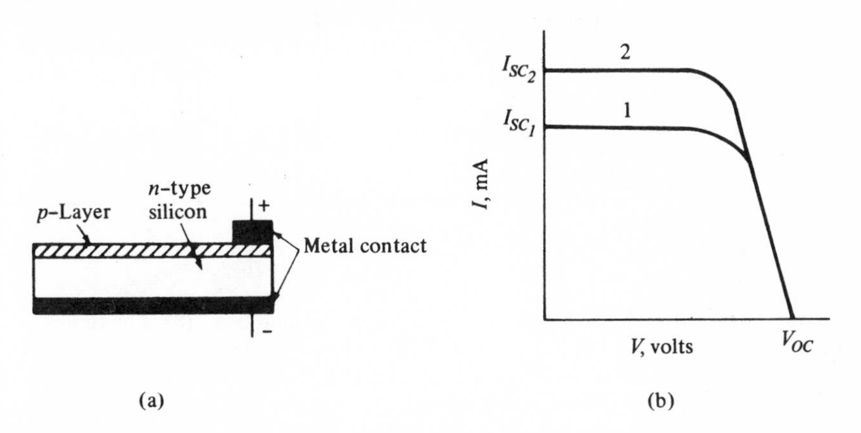

Figure 10–8 (a) Cross section of silicon solar cell. (b) *I-V* for solar cell as illumination is increased.

The spectral response for photovoltaic devices ranges from 5000 Å for selenium to 50,000 Å for indium antimonide. The latter response is above the visible range and well into the infrared spectrum. For this reason, this type of material is used for infrared detection and not for power development.

Finally, silicon cells respond much faster than other photodevices. Response time of a few microseconds is common. For this reason silicon cells are also used in computer hardware, such as card readers.

10-4 Photodiodes and Phototransistors

Earlier in the text it was revealed that small amounts of current flow when a diode is reverse-biased. It was also mentioned that when heat energy was added, this reverse current increased. The added energy was sufficient to break a valence-bond electron from its parent atom and raise it to the conduction band of energy. In photodiodes the same technique is used except that heat energy is replaced by light energy. The light is focused on a *p-n* junction with the aid of a lens, as shown in Figure 10-9. In Figure 10-9(a) a point-contact diode is shown and in Figure 10-9(b) a junction type is shown. In the point-contact photodiode the outside metal shell serves as a cathode; in the junction type, the *n* material is the cathode. Figure 10-10(a) shows typical characteristics of a photodiode. The results are somewhat peculiar and require an explanation. In the fourth quadrant the photodiode acts like

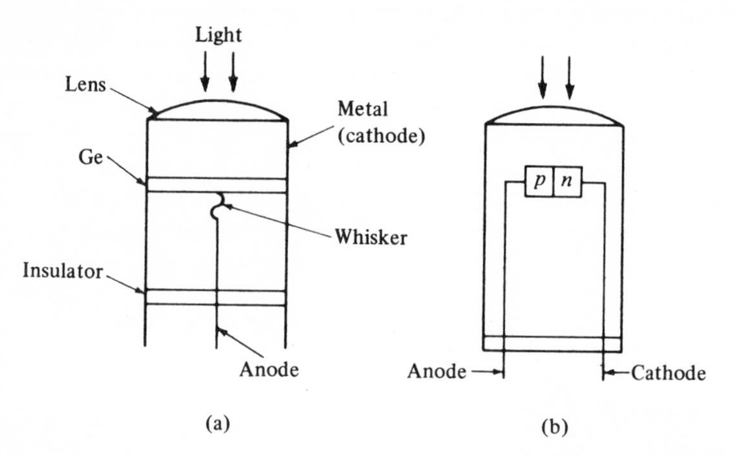

Figure 10–9 (a) Cross section of whisker-type photodiode. (b) Junction-type photodiode.

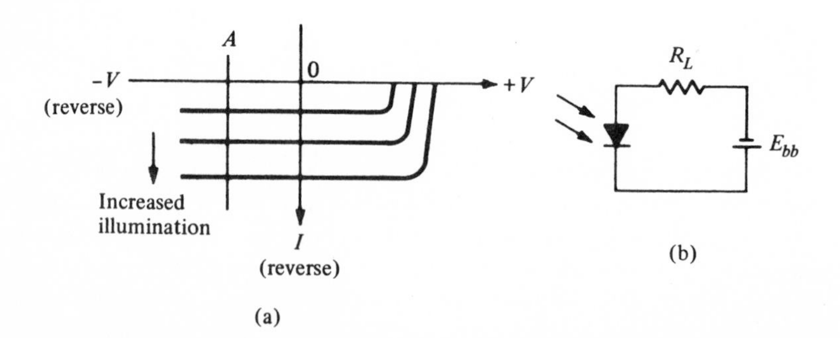

Figure 10–10 (a) Typical photodiode *I-V* response. (b) Simple photodiode circuit with reverse bias.

a photovoltaic device; that is, it produces a voltage. Notice that the curves are similar to the silicon solar cell of Figure 10-8 except that the quadrants are different. Quadrant three shows how the photodiode behaves when reverse voltage is *applied.* In quadrant four a voltage was *developed.* If we keep a constant voltage in quadrant three, line *A* in the figure, the current increases with increased illumination. This produces a lowering of resistance across the diode. Therefore, if we reiterate the cause and effect, increased illumination produces decreased resistance, we have a description of a photoconductive cell. We can say that a photodiode can be used as a photovoltaic or a photoconductive device.

Figure 10-10(b) shows a simple circuit in which the photodiode is used as a photoresistance. With no light present the current that flows is the typical current for an ordinary reverse-bias diode.

Photodiodes have a wavelength spectrum from visible light to and including the infrared region. The output current is relatively low and therefore requires immediate amplification. Putting it simply, the output current cannot energize a relay and therefore the diode output would require a stage of amplification between the photodiode and the relay. However, the response time is relatively fast and makes the photodiode useful in high-speed counting and in soundtrack pickups.

A phototransistor behaves just like a photodiode except that it permits us to reap the added benefits of current amplification. The light is focused through a lens on the base–emitter junction. Figure 10-11(a) is a diagram of the phototransistor and Figure 10-11(b) shows the symbols. Notice that only emitter and collector leads are necessary. The operation is quite similar to a CE amplifier. Light striking the transistor causes a base current to flow. The resulting collector current is greater than the light-stimulated base current by a factor of β. Figure 10-11(c) shows a simple phototransistor

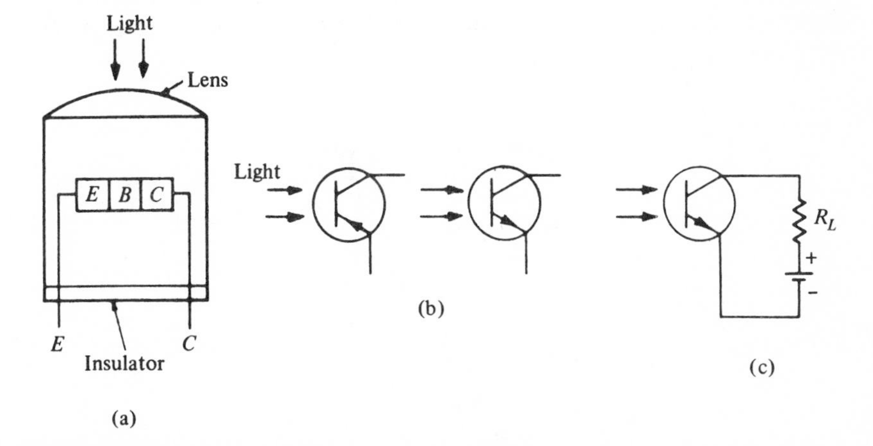

Figure 10–11 Phototransistor. (a) Cross section. (b) Symbols. (c) Circuit.

circuit. It is quite obvious that the output current will be greater in a phototransistor than in a photodiode for the same illumination. Typically, it has a current gain of 100 to 500. However, on the negative side, a phototransistor has a slower response than a photodiode. Spectrally, a germanium phototransistor peaks at about 16,000 Å.

In some phototransistors the base lead is brought out for external connections. In applications in which the light level is low, a slight forward bias may be used to assist the light-induced current flow.

10-5 The Light-Activated SCR (LAS)

One can readily imagine how a light-activated SCR functions if the operation of SCRs is understood. A previous chapter showed that SCRs are capable of conducting current if the gate current is adequate. Once the gate permitted current flow, it lost control. To regain control, the anode voltage had to be reduced to zero or a negative voltage applied. The light-activated SCR operates similarly except the gate current depends on intensity of light. If sufficient light is present, the SCR will fire. The light source loses control until the anode voltage is reduced to zero. Figure 10-12(a) shows a simple dc circuit wherein a light source, say sunlight, will fire the LAS. The only way to turn it off is to interrupt the anode circuit with switch *S*.

The mechanical interrupting process is indeed impractical in some control circuits. It was pointed out in the SCR chapter that the interrupting can be done by using ac in place of dc. Figure 10-12(b) is the same circuit except that an ac source is used. Current will flow through the LAS and lamp every half-cycle when point *A* is positive with respect to point *B* and when sufficient light is present to trigger the SCR. If the light intensity drops below trigger level, the LAS will cease conducting as soon as the ac voltage causes point *A* to become negative with respect to *B*. Hence, the LAS circuit will remain

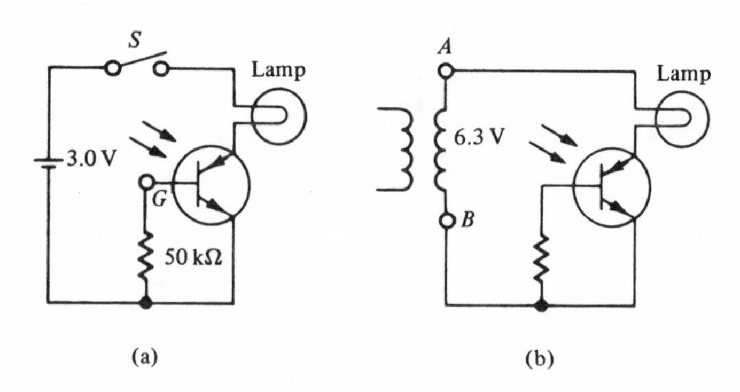

Figure 10–12 Light-activated SCR using (a) dc and (b) ac.

dormant until there is again sufficient light intensity to cause anode current to flow.

Figure 10-12(a) can be used in opening a garage door if we replace the lamp with a relay. Once the headlights strike the LAS, the LAS passes enough current to energize the relay, which initiates a motor to lift the door. Obviously, ac power to the LAS would not be appropriate because once the car light beams lost "contact," the relay would be deenergized and the door lifting would cease. (Overriding circuits can be used to correct this condition.)

Figure 10-12(b), on the other hand, could be used where a counting process is required. If the lamp is replaced by an electric counter, light interruptions can be recorded because individual light bursts would cause current flow. However, there is a limit to how fast the counting process may occur. LAS as well as other electronic devices have upper-frequency limitations. Simply putting it, the LAS requires a finite amount of time to start to respond when light strikes and a finite amount of time to "unrespond" when light is removed. If the light bursts come much faster than these responding times, the LAS will be on all the time and therefore cease to count usefully. This is like the drill sergeant barking out orders faster than the rookie can carry them out. Once again, we repeat that each electron device has its own upper-frequency limit.

10-6 Summary

In this chapter we discovered that light is wavelike in nature. It travels through space just like high-frequency radio energy. The frequency associated with light is well above the radio-frequency limit. Light phenomena are generally identified in terms of wavelength instead of frequency. Mathematically, this relationship is

$$\lambda = \frac{V}{f}$$

Each visible wavelength produces a different color response in our brain via the eyes. The most sensitive colors are in the yellow-green to yellow-red region. Immediately outside the visible range are two important regions, called infrared and ultraviolet. These two ranges, and the visible range, are sensed by solid-state photodevices. The range of wavelengths that are sensed depends on the material doing the sensing. Just like the eyes, each device favors its own range of wavelengths.

These solid-state photodevices are classified as photoconductive or photovoltaic. Photoconductive devices effectively lower resistance with light illumination and photovoltaic devices produce a voltage. Generally, photoconductive devices act as transducers, converting a light source to an electrical signal. A typical application would be a mechanism for opening a garage door when the headlight beam strikes the photoconductive device. Of course, the minute photoresponding current has to be amplified before useful work can be effected.

A photovoltaic example is the use of silicon solar cells to power satellites in space. The individual cell produces low power but when connected in series and parallel, a group of cells can produce reasonable power.

Of special interest are the photodiode and phototransistor. The photodiode shows characteristics of a photovoltaic and a photoconductive device. In the reverse-bias application it acts as a photoconductive device. The phototransistor behaves just like the reverse-bias photodiode except that an extra advantage of current gain is realized. In both devices the principle of operation is the focusing of light energy on a diode. Just as heat energy causes more current, so does light energy break a covalent bond for conduction.

Questions and Problems

1. Convert the following wavelengths to meters: (a) 0.6 μm, (b) 32 μm, (c) 0.08 μm.
2. Convert the following wavelengths, given in meters, to micrometers: (a) 14×10^{-7}, (b) 0.82×10^{-5}, (c) 16.2×10^{-6}.
3. Convert the following frequencies to wavelengths in meters: (a) 150 kHz, (b) 150×10^{6} Hz, (c) 150×10^{12} Hz.
4. Which of the three frequencies in Problem 3 is nearest to the light region?
5. Convert the answers in Problem 3 to micrometers.
6. Convert the answers in Problem 3 to angstroms.
7. The frequency range for AM radio broadcasting is from 500 to 1500 kHz. Convert this to a wavelength range in meters per cycle.
8. Convert the visible range of wavelength in Figure 10-2 to a frequency range.
9. Explain how a photoconductive device operates.
10. Two factors that describe a photoconductive device are (1) resistance dark-to-light ratio, and (2) wavelength response. Explain these.
11. How does a photovoltaic device work?
12. What is meant by the efficiency of a photovoltaic device?
13. What makes a photodiode a unique device?
14. What advantage does a phototransistor have over a photodiode?
15. What principle makes the photodiode and transistor produce carriers?

11

Integrated Circuits (ICs)

11-1 Advantages of ICs

The early history of electronics involved vacuum tubes such as diodes and triodes. For many years these devices were extensively applied and reapplied as newer variations were developed. When these applications seemed to reach a plateau, a new device appeared (1948) on the market—namely, the transistor. The vacuum tube was not only challenged, but was actually replaced in many applications. Moreover, the transistor offered so many new advantages that it opened new areas of electronics. Then by the time the transistor could reach its plateau a new "device" had made its appearance and opened new vistas of applications. The new device is the integrated circuit. Perhaps the IC should be more appropriately called a new technique because the electronic components are still the same except that they are smaller and the circuit is manufactured as a unit instead of as individual components.

The definition of an IC may vary depending on the user of the words. IEEE defines it as "a combination of interconnected circuit elements inseparably associated on or within a continuous substrate." In simple terms, ICs are resistors, capacitors, and transistors that are miniaturized and interconnected to make a complete circuit or part of a complete circuit. The space requirement to achieve this is no larger than that required by an ordinary transistor.

In all probability, this is the first exposure for the student to the field of ICs. It is, therefore, appropriate to give an overview of this field before specifics are discussed. One appropriate question that, if answered, will give some insight is "Why ICs?"

Over the years miniaturization has been sought. In recent years weight reduction in satellites and missiles has been desired so that payloads may be increased. If a complete amplifier can be obtained in a package equal in size to a transistor, then indeed weight reduction is achieved if many of the circuits are in IC form. In addition, computer experts have been attempting to increase the speed of modern-day computers. The speed, in this case, is the rate at which a computer does a given operation. If the computer physically is long then it takes a finite amount of time for a bit of information to go from one point to another. At 186,000 miles per second, it takes one nanosecond (1×10^{-9} second) for a signal to travel one foot. For computer men this is sometimes considered slow. Hence, to decrease travel time, components must be placed closer together, and the only way to decrease distance is to use ICs.

Besides the above two reasons for IC, there is another. In the manufacture of individual components, it is a known fact that a certain percentage, even though small, will be defective. Since one hundred percent inspection is not usually the rule, these individual defective components will find themselves in a given electronic device. If many such parts make up the electronic system, it is obvious that the chances of producing a malfunctioning system are greater. In technical terms, the reliability decreases as the number of parts increases. It is common today for a person to buy a car or TV or dishwasher and find it inoperable or malfunctioning the first time it is used. This of course is a "chance" one takes. Manufacturers can decrease your chances of getting a "lemon" by inspecting each component. This contributes to the cost considerably, and so a compromise is reached. Similarly, in electronics many components with a very low defect rate make up a system. A particular system will fail if the chances of defect increase. In addition, each component must be soldered or wire-wrapped, and here is another source of defects and therefore a decrease in reliability. In ICs, components are made and wired practically simultaneously in one package. This single package is equivalent in reliability to a single component in the conventional approach of individual parts. Therefore, if an IC package, which is a complete system, approaches the reliability of an individual component a distinct advantage is gained by using ICs.

Besides the improved reliability advantage, ICs have a cost advantage in packaging. In the conventional approach, each component, resistor, capacitor, or transistor has to be packaged. In ICs, only one package is necessary. In addition, the IC process produces hundreds of circuits simultaneously. This implies that perhaps thousands of components are also produced. So the technique of mass production is not abandoned with the new technique of integrated circuits.

One would imagine that with the advantage of ICs, circuit designers would turn to them en masse. Indeed, it is evident that the full potential of these circuits has not been realized since IC manufacturers predict substantial increases in sales. However, because of the relative newness of these devices, application engineers are still in the process of reorienting their thinking from discrete components to ICs. In addition, manufacturing costs are still relatively high. This of course is typical of a new product; but as experience is gained, processing will stabilize and thus costs are predicted to decrease.

The plateau for ICs is some years ahead. The use of these devices will increase manyfold as more application engineers become more familiar with them and as processing techniques improve. Progress is made in application and production in what seems a daily fashion. As an example, ICs are presently considered low-power devices because of their limited power-dissipation ability. It is assumed that process engineers have their attention directed to improving the power-handling ability of ICs and will undoubtedly succeed. This is precisely the pattern ordinary transistors went through: first a small-signal device and then, after considerable research, high-power devices. Hence, as we proceed through this chapter, limitations that are mentioned may no longer exist because of this rapid rise to a possible technological IC plateau.

11-2 Semiconductor ICs

There are two general classifications of integrated circuits: semiconductor monolithic ICs and thin-film ICs. The word monolithic means "single stone." Therefore, an IC circuit that appears on a single silicon crystal is considered a monolithic IC. The fact that the "stone" or substrate is semiconductor material further identifies it. When the substrate is an insulating material, such as glass, the IC falls into the category of thin films. In both methods, miniaturized transistors, capacitors, diodes, and resistors are possible. Each approach has its own advantages and these will be investigated as we proceed through the chapter.

Each technique has its limitations, and engineers have come up with a marriage of thin-film and semiconductor ICs to produce a third classification, called a hybrid IC. This type is used where special requirements must be met, but in general it is more expensive than the other two.

Before we proceed too far in semiconductor ICs, it is important to have a feeling of the physical size of these circuits. A given silicon wafer may be one inch in diameter. By the time the IC process is completed, 200 to 1000 identical circuits may be fabricated. These are precisely cut into individual parts, the leads are attached, and the circuits are tested and packaged. All of this must be done under a microscope because of the small size of each circuit. In terms of numbers, a conventional resistor can be reduced by a factor of 100,000 when duplicated in an IC and the completed circuit with its associated transistors, diodes, resistors and capacitors could be contained in a letter "O" of this size.

The basic process of ICs is a duplication of making ordinary silicon transistors. In this process a block of semiconductor material is exposed to two selected diffusion areas. If we start with a *p*-type material, and subsequent diffusions are *n* and then *p*, a *p-n-p* transistor results, with the original *p* material serving as the collector. In IC fabrication, other areas in the crystal are *simultaneously* diffused to produce resistors, capacitors and, diodes. A resistor is made of either *p* or *n* material; a diode is formed by two diffused regions; and a capacitor is similar to a diode except that it requires a greater area. Therefore, it seems as if the transistor is the most complicated component in semiconductor IC manufacturing.

Let us go through the process of diffusing a transistor, keeping in mind that it is only one component of many and only one transistor of perhaps a thousand on a one-inch silicon wafer. The basic building block is the silicon wafer upon which a fine layer of semiconductor material is grown. The starting wafer is the substrate, which offers mechanical strength; the thin epitaxial layer is the region in which the transistor is diffused. Both regions may be *n*-doped or *p*-doped, but the thin upper layer has a higher resistivity. In the formation of a transistor, the epitaxial layer may be omitted; however, it does provide certain electrical advantages. When this wafer is exposed to an oxygen atmosphere a thin, insulating oxide layer is formed. This layer protects the wafer from contamination. Figure 11-1 shows the three basic starting layers. Keep in mind that the whole wafer is so prepared, and that masks are made with tiny holes that permit the removal of the insulating SiO_2 layer. We will not investigate the photoresistor process in detail; let it be sufficient to say that these masks are prepared with appropriate windows. (Essentially the mask is made from a large drawing and then reduced in size photographically to fit the wafer.) The windows permit selected regions to

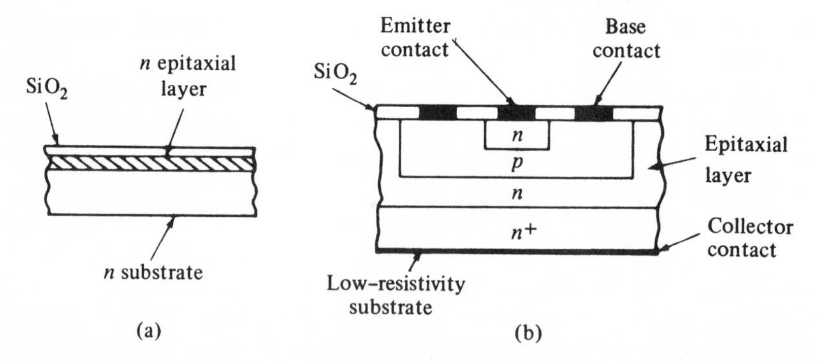

Figure 11–1 (a) Cross section of silicon wafer with appropriate layers. (b) Cross section of a diffused discrete transistor.

be exposed for removal of SiO_2 and subsequent diffusion of the exposed region by appropriate dopants. If a thousand windows were etched in the SiO_2, a thousand *n*-epitaxial areas would be exposed for doping with a *p*-forming substance. With proper temperature and time of exposure, the diffusion depth can be controlled. We now have the base (*p* type) of a transistor. The original *n*-type epitaxial layer serves as a collector. Another masking and diffusion process follows. In this case the *n* dopant is of such concentration that it cancels some of the *p*-type base and saturates a small portion of the base to an *n*-type region. Again, the temperature and time control the depth of diffusion. This new region serves as the emitter. Figure 11-1(b) shows these regions. Another layer of SiO_2 is grown on top to protect the base–emitter junction. Finally, this SiO_2 is selectively etched to permit contact to the emitter and base. A thin metallic film is vaporized over the total surface, but then removed from the SiO_2 and permitted to remain at the SiO_2 windows or emitter and base. A collector contact is made at the bottom by way of a metal contact.

If a thousand of these transistors were made on a single wafer, a cutting process would yield a thousand discrete transistors. Notice that in the wafer form all the collectors were made common by the substrate. Indeed, if these same transistors were to be connected to other surface-diffused components, as in ICs, then the collector contact must be brought out to the surface for convenient connecting. Moreover, the substrate is common to other diffused components or devices, and therefore isolation must be carried out. In other words, alongside the diffused transistor may be a diffused diode in which the *p*-type substrate could be one portion of the diode and an *n*-diffused region above it the other. Hence, the collector of the transistor and the anode of the diode would be tied together because they share a common substrate. In some cases this may be desirable, but it is conceivable that flexibility would be lost if ICs were manufactured this way. Fortunately, engineers have managed to isolate the collector, and thus the process is not too different from discrete-transistor manufacturing. Figure 11-2 shows the basic dif-

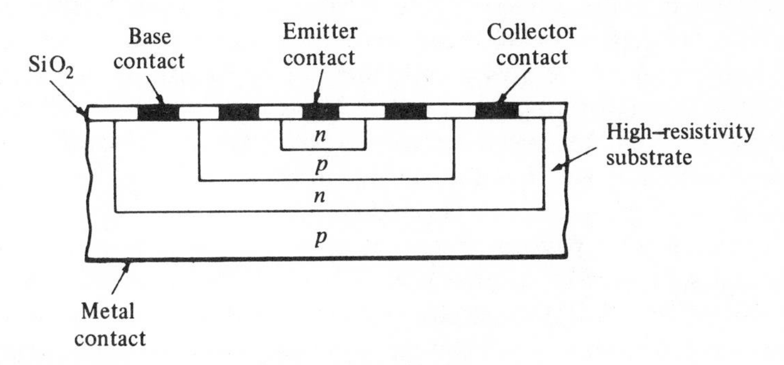

Figure 11–2 Cross section of a diffused IC transistor.

ference. In this case an *n-p-n* transistor is made into a *p* substrate. (If discrete transistors were being manufactured, an *n-p-n* transistor would use an *n*-type substrate.) The isolation is achieved by placing a reverse bias between the *p* substrate and the *n*-type collector. This, of course, produces a high resistance and even if another component shares the substrate this relatively high resistance appears between them.

With such intense involvement in the IC process for a transistor, we must not forget that diodes, capacitors, and resistors are made simultaneously with transistors. Diodes and capacitors are made during collector-base or base-emitter diffusions and resistors are made during the base region diffusion. Figure 11-3 shows all three components, along with the transistor for comparison. Notice that the "collector" region in the resistance, Figure 11-3(c), is covered with SiO_2 and that the "emitter" in the diode is missing. This, of course, is achieved by selective masking, etching, and diffusing. In all three figures the substrate is not shown. To interconnect components, metalizing is achieved on top of the SiO_2 layer. Masking techniques are used to define the geometries.

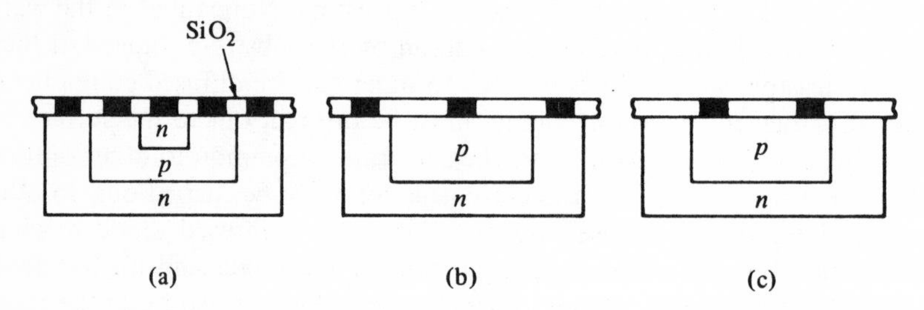

Figure 11–3 Cross section of an IC (a) transistor, (b) capacitor or diode, and (c) resistor.

It is worth mentioning that a *p-n* junction type capacitor has its drawbacks; for example, it is voltage-sensitive. One approach for circumventing this problem is to make a different type of capacitor. This can be done by using the SiO_2 protective sheet as the dielectric. It does not take a great stretch of the imagination to see how a capacitor can be formed. If the SiO_2 is the dielectric, a metalized region on top of the SiO_2 can serve as one plate and a heavily doped emitter region can serve as the other conductor or plate. This type of capacitor also has the advantage of not being polarized; that is, voltage of either polarity can be applied.

To complete the physical picture of interconnecting and component profiles in ICs, Figure 11-4 is introduced. Figure 11-4(a) is the top view, 11-4(b) the profile, and 11-4(c) the electric circuit. Keep in mind that these are only three components on a small section of the wafer and that many, many more can be, and are, placed and interconnected before the wafer is subdivided into perhaps a thousand of these completed circuits.

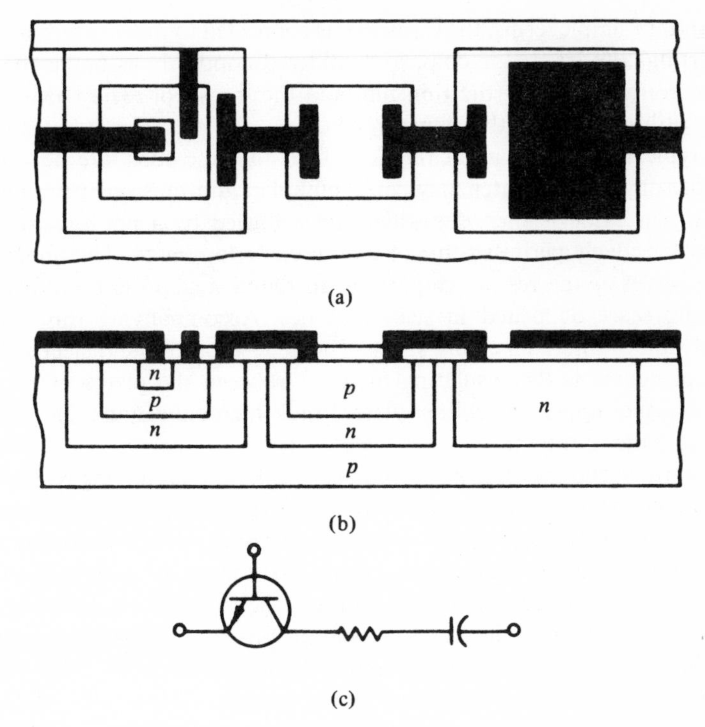

Figure 11–4 An interconnected IC. (a) Top view. (b) Side view. (c) Circuit-diagram.

11-3 Thin-Film ICs

Usually an alternative process is devised in any system because it has some advantages over the original process. Thin-film ICs have one obvious advantage because the substrate is ceramic or glass. Therefore, isolation between components is not a problem, as it is in semiconductor ICs.

Basically, techniques such as evaporation and sputtering form a pattern of passive elements (resistors and capacitors) on the substrate. After these are on, active elements, such as transistors, are added separately. At present, depositing a transistor simultaneously with the other components is not quite feasible. The process of separately adding active devices makes the thin-film IC more expensive than the semiconductor IC. The addition of the active device to the thin film is done simply by interconnecting leads from the transistor or diode chip to the proper places in the thin film. Since the chip is brazed to a prepared area on the thin film, only one additional lead connection is necessary for a diode and two for a transistor. A more clever approach for connecting thin-film passive elements to active elements has been devised by the IBM Corporation. This technique is called the flip-chip method. In simple terms, the transistor is connected to the thin film by flipping (inverting) the transistor chip, and soldered copper balls make the appropriate contact between the thin film and the transistor. The IBM-360 computer utilizes flip-chip thin-film circuits.

A typical thin-film circuit requires, first, that the substrate be very clean because any contaminates may cause pinholes, and thus open circuits would result. Tin oxide is then deposited and followed by a photoresist process, which selectively permits some of the tin oxide to remain. The remaining tin oxide serves as the resistance pattern. In the next step the bottom plates of capacitors are deposited in selected areas. Accompanying this may be a deposition of interconnecting strips. The deposition of the dielectric follows. This of course is the insulating material between the plates of capacitors. Finally, the upper capacitor plates and interconnections are added to complete the thin film.

A word or two may be appropriate on the science of making resistance and capacitance in thin-film circuits. Most of the deposited thin films range in thickness from 100 to 10,000 Å. As a comparison, a single atom has a diameter of approximately 2 Å. So we can see that the layers are indeed small. To effect a deposit on an insulating substrate, the clean substrate is placed in a vacuum chamber and heated in the atmosphere of the desired evaporant. Heating the substrate drives off any surface contaminants. The value of a deposited resistance is determined by the relation

$$R = \rho \frac{l}{tw} \tag{11-1}$$

where ρ is the resistivity, t is the thickness, w is the width, and l is the length of the deposited layer.

If we group ρ/t and call it sheet resistance, Equation (11-1) becomes

$$R = R_s \frac{l}{w} \tag{11-2}$$

where R_s is quoted as ohms per square. A typical value for R_s is 200 ohms per square. The unique feature of Equation (11-2) is that resistance depends on shape rather than on size. As an example, if l/w is $3/1$ for a large configuration, it would produce the same resistance for a much smaller configuration as long as the l/w ratio is $3/1$ and the same material is used. Of course, there is a limit to how small a given resistance can be made. Power dissipation is one consideration.

Figure 11-5(a) shows a typical deposited film resistance. The resistor is made from nichrome, the contacts are silver or aluminum, and the protective oxide is silicon monoxide.

For capacitance film deposition, the appropriate equation is

$$C = k\frac{A}{d} \times 8.85 \times 10^{-12} \tag{11-3}$$

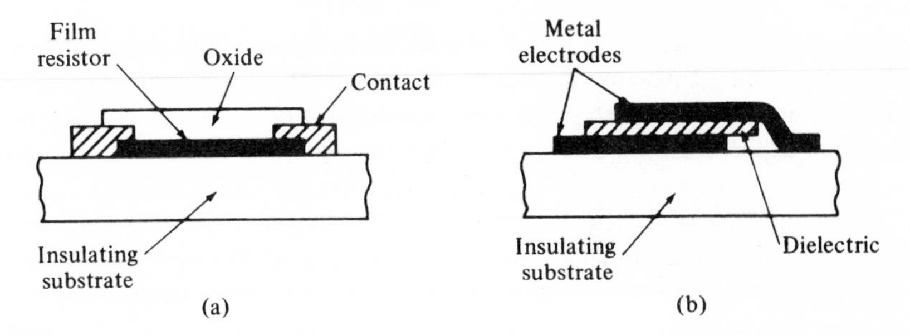

Figure 11–5 Thin-film (a) resistor and (b) capacitor.

where k is the dielectric constant, d is the dielectric thickness, and A is the electrode area.

The silicon monoxide that is used as a cover for resistors can be used as the dielectric. Its k value is approximately 4. The formula indicates that the capacitance value can be increased by increasing the k value. Tantulum oxide has a value 4 times higher. Other substances, such as titanium oxide, increases the k, and hence the capacitance, even further. The dielectric constant is important because it represents a means of control of the magnitude of capacitance other than the area and distance between plates. Figure 11-5(b) shows a typical thin-film capacitor.

No mention has been made so far of a third passive element, the inductor. Although these elements are made, the limit of inductance is quite restricted (to the microhenry range). Because of this limitation and manufacturing difficulties, film inductors are not used extensively. This restriction exists today but, as mentioned earlier, techniques are devised almost on a daily basis that overcome this type of problem. The same statement holds true for active devices, such as the transistor. Constant pressures on this problem at many places will produce practical thin-film transistors.

11-4 Hybrid ICs

Earlier it was mentioned that when a thin-film or semiconductor IC cannot do a particular job, a combination, or hybrid, IC is used. One of these factors may be a high voltage requirement. In any case, a hybrid IC can take on different forms. The first of these is the *multichip*. The multichip hybrid IC is composed of several chips or discrete monolithic blocks (film or semiconductor) within a single package. These parts are attached to a ceramic substrate and then interconnected. By this approach, isolation—which is a problem in monolithic circuits—becomes less of a problem. Indeed, some flexibility is introduced, because each chip can be electrically optimized; that is, compromises can be avoided. One example is the production of *n-p-n* and *p-n-p* transistors on the same semiconductor substrate. This is quite difficult.

Also, another advantage is achieved because failure or a redesign of one chip does not require throwing the whole IC out, just the affected component. On the negative side, most interconnections are done one at a time. This proves to be quite expensive. In addition, for every connection a decrease in reliability occurs.

A second type of hybrid IC is the *compatible* circuit. In this type, thin-film components are used in combination with semiconductor ICs. The basic circuit has a semiconductor IC circuit, upon which a thin insulating layer of SiO_2 is grown. On top of this SiO_2 substrate, thin-film components are formed. These components, in turn, are joined to the semiconductor IC through windows in the SiO_2. The chief advantage here is that the more critical components are fabricated by the superior thin-film method. This advantage will be discussed in a subsequent section. It is also worth noting that this process requires more materials and steps in manufacturing. Each of these contributes to a lowering of product reliability and increase in cost as compared with the monolithic semiconductor IC.

A third class of hybrid IC is the *monobrid* which is a contraction of monolithic and hybrid. Some circuits can be made on a single substrate but are too complicated. The complication may require a larger chip for many more components. This complication causes a lower yield and therefore a higher cost. To get around this problem the circuit is split into two parts, each of which is a monolithic circuit. The two are joined with wire bands and placed in a single package. With this approach some optimizing can be achieved, such as that which may be gained by separating *n-p-n* and *p-n-p* transistors. Two or more monolithic circuits comprise a monobrid. However, as the number increases we approach the disadvantages of a multichip hybrid—that is, a decrease in reliability and an increase in cost.

11-5 Limitations

Some of the advantages and disadvantages of the three types of IC were mentioned in Section 11-4. First of all, in the monolithic semiconductor IC the range of passive component values is rather limited. This is so because the resistivity of the material used depends on simultaneously diffused transistor elements (emitter, base, and collector) and, say, a resistor. A compromise is reached so that the transistor characteristics do not suffer too much. High resistivity, which is required for the resistor, adversely affects the transistor. Or, a large value of resistance or capacitance requires a large chip area, which indeed is a physical limitation. Second, because of process capability, close tolerances on passive elements are difficult. A typical tolerance is ± 20 percent. Third, a semiconductor IC has high-frequency limitations because of parasitic capacitances that appear between diffused components and substrate. Finally, the cost of these ICs is relatively small provided they are manufactured in large quantities. However, if a small number is desired the cost per unit is exorbitant, primarily because of high mask costs. These costs much be included in the price whether many or only a few ICs are made.

One advantage the thin-film ICs have over the semiconductor IC is that the range of values for passive elements is wider. In addition, a closer tolerance is possible. On the other hand, thin-film circuits are larger and more expensive, primarily because of the need to "reserve" space for an active element such as a transistor. The space requirement is dictated by the physical size of a discrete transistor and bonding space. With all this room, a semiconductor IC could diffuse a relatively large complete circuit. Because of active component addition and subsequent bonding, the cost for thin-film circuits is relatively high. They become competitive only when the number of active components is small.

The hybrid IC has the distinct advantage of adding flexibility to ICs. The advantages of both types of circuit are combined in one system. The tolerances of passive components in thin films can be used with the optimized transistor characteristics of a semiconductor IC. However, the hybrid has the disadvantage of requiring interconnections between parts. This contributes to a lower yield and therefore a higher price.

11-6 Summary

In this chapter we briefly scratched the surface of a new technology—namely, integrated circuits. The chief advantages of ICs are reduction in size and weight, increase in reliability, and a potential cost saving over conventional components.

An integrated circuit starts with one of two basic building blocks. If the block or substrate is a semiconductor, then it is classified as a semiconductor IC. If the substrate is an insulator, and a vaporization process produces components, it is a thin-film IC. In both cases, components are diffused or vaporized through masks. The amount of diffusion and vaporization is controlled and directed through holes or windows in the masks. A photographic process is required to reduce the mask to such a size that perhaps 800 complete circuits can be made on a one-inch-diameter silicon wafer. The interconnections between components are also masked; this practice incidentally improves reliability because the human factor is removed from the process.

In semiconductor ICs, the transistor is the most complicated component to make. If the epitaxial layer is *n*-type, a second diffusion by *p* material in selected areas can convert some of the *n* region to *p*-type. The epitaxial layer becomes the collector and the new diffused region the base. Another diffusion, this time *n*-type material, on top of the base will produce an emitter. The window for the emitter must be smaller than the base region. The penetration is controlled to produce an *n-p-n* transistor. It should be obvious now why semiconductor ICs have resistor limitations. The resistivity for the emitter is such that the *n-p-n* transistor characteristics are satisfied. However, in some other region in the wafer, a resistor is simultaneously made. Its resistance value is limited by the chosen emitter resistivity. Even with compromises, limitations are always present.

Thin-film ICs have a wider range and better tolerances for passive components than semiconductor IC components. Some active components such as transistors are difficult to manufacture. It is for this reason that a marriage was made between thin-film and semiconductor ICs, resulting in a hybrid class of ICs. The hybrid IC requires more interconnections and, therefore, has a reduced yield and higher cost. In any case, where cost is no problem the hybrid IC can do well where the other two cannot.

Questions and Problems

1. What is an integrated circuit?
2. What are the three types of ICs?
3. What are the advantages of ICs?
4. How does a computer benefit from IC circuits?
5. What is meant by reliability?
6. What are the basic three layers of an IC wafer?
7. What process permits diffusion of a semiconductor IC in selected areas?
8. What is the basic difference between a semiconductor IC and a thin-film IC?
9. What disadvantage does the semiconductor IC have relative to its substrate?
10. How is the depth of diffusion controlled in a semiconductor IC?
11. In thin films, what is the name of the process whereby a layer of resistance is deposited?
12. What advantage does the thin-film resistor have over the semiconductor resistor?
13. Is the capacitor in Figure 11-4 a reverse-bias type? Explain.
14. What are the three types of hybrid ICs?
15. (True or false.) Hybrid ICs are generally cheaper than other types of ICs.
16. (True or false.) It is easy to make *n-p-n* and *p-n-p* transistors on the same chip.

17. Starting with a simple wafer of silicon, add the necessary layers and describe how discrete transistors are made.
18. (True or false.) Inductors are not common in ICs.
19. How is isolation between components achieved in ICs?
20. What is a flip-chip?
21. (True or false.) SiO_2 is a conductor.
22. (True or false.) In semiconductor ICs a resistance is made when the base of a transistor is made.
23. (True or false.) In semiconductor ICs a diode is similar to a capacitor.

12

Other Solid-State Devices of Interest

12-1 Zener Diodes

Once transistors had established themselves as useful and reliable electronic devices, further research and development produced many other solid-state devices. Many of these replaced vacuum tubes; others opened new horizons in electronics. Many of these are used quite frequently in circuits; others are used infrequently because of their limited application. And indeed many are relatively new to the applications people and hence have not been fully exploited. The source of these special solid-state devices seems unlimited.

This chapter will investigate a few of these devices. Some are quite common at this date; others are just being investigated.

The first of these, which has already established itself in electronics, is the Zener diode. An earlier chapter on diodes revealed that there are two parts to a diode characteristic: the forward *I-V* response and the reverse *I-V* response. We concentrated our attention on the forward characteristic, wherein a slight increase in voltage yielded considerable current. It was also mentioned that in the reverse direction many volts produced only microamperes and that, in any rectifier application, we avoided high voltage for fear of breaking down the diode. It is this breakdown region that we make use of in Zener diodes. Figure 12-1 shows a typical *Zener-diode characteristic*; our attention here is directed primarily toward the reverse breakdown region. This steep rise tells us that, in this region, the voltage is relatively constant

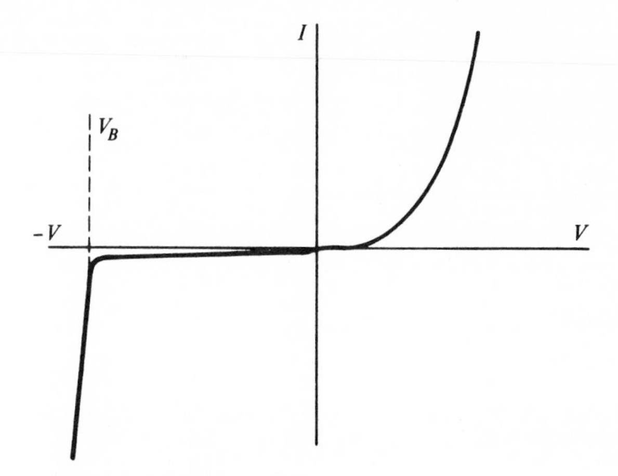

Figure 12–1 Zener-diode characteristic.

for various amounts of current. This is quite unlike ordinary active devices, in which a change in current is reflected in a corresponding change in voltage. It is precisely this feature of the Zener diode that is exploited in many ways.

One common use is in a voltage regulator. As a preliminary bit of information, voltage regulators are employed where a steady dc voltage is required. A constant dc voltage is not easily obtained, either because the input fluctuates or because the load demand changes. Both cases result in output voltage changes, which in some electronic equipment are undesirable.

To offset large changes in output voltage, Zener diodes are used, as shown in Figure 12-2. The unregulated voltage is the ouput of the generator. It can be assumed that it fluctuates, but that a steady voltage is desired across R_L, which is also the voltage across the Zener diode. Therefore, if the Zener diode is in the breakdown region, where the voltage is relatively constant, the output voltage across R_L will remain relatively constant. In other words, changes in the generator voltage will appear as much smaller changes across the load because of the Zener diode.

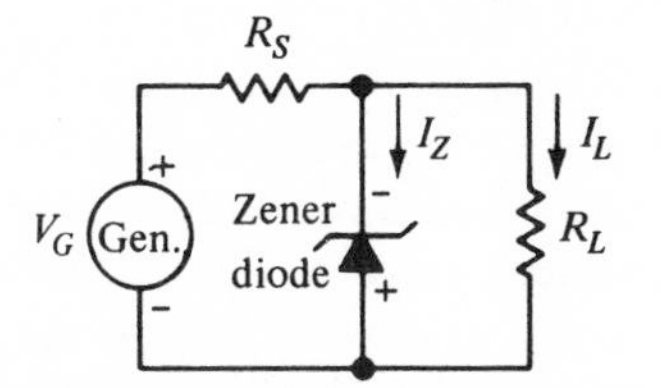

Figure 12–2 Zener diode in a regulating circuit.

The foregoing action will occur if the Zener diode is in the breakdown region. This of course depends on the generator voltage, R_S, and the current through R_S. In most cases R_S is the unknown factor, and looking at Figure

12-2 we can obtain the relationship

$$R_S = \frac{V_G - V_L}{I_Z + I_L} \qquad (12\text{-}1)$$

where V_L is the desired regulated load voltage, I_L is the maximum load current, and I_Z is arbitrarily chosen as 10 percent of the maximum current.

EXAMPLE 1

A generator that fluctuates puts out a maximum of 50 volts to a load that requires a maximum of 30 mA at 10 volts. Determine the values of R_S necessary for a 10-volt Zener diode.

SOLUTION

$$R_S = \frac{V_G - V_L}{I_Z + I_L}$$

$$= \frac{50 - 10}{(30 + 0.1 \times 30)10^{-3}}$$

$$= 1200 \text{ ohms}$$

The convenient part about Zener diodes is that they have breakdown voltages from approximately 3 to 200 volts and can be placed in series to produce that many more breakdowns. This point may seem insignificant, but gas tubes used as regulators have always lacked the extra flexibility of numerous breakdowns. This point is covered in Chapter 13.

Along with a voltage rating, Zeners have maximum current and power ratings. Under steady-state conditions these quantities cannot be exceeded. Any application must be investigated before energizing or if energizing is the only way out, these values must be monitored and compared to manufacturers' limitations.

Returning to Figure 12-2, notice the polarity of source and Zener. For Zener action the diode must be in reverse bias. It may be asked why it is that the static characteristic takes on a vertical response at a particular voltage. The answer is that with controlled and heavy doping of a diode a Zener increases the chances of an electron breaking away in the reverse-bias direction. Collisions between free electrons and neutral atoms occur, further releasing more carriers. At a critical voltage, called the Zener breakdown, this action is quite instantaneous and suddenly many electrons are available. Now the voltage is quite constant and a large number of carriers are available for conduction.

12-2 The *p-i-n* Diode

One of the drawbacks of ordinary transistors is that they are not high-frequency devices, like vacuum-tube klystrons or magnetrons. Transistors are capable of producing oscillations at frequencies of 2 GHz, or even higher, but these devices are quite expensive to manufacture. In recent years development of microwave diodes led to the operation of these devices at more than 50 GHz. These diodes, which are only two-element devices, require only dc for operation. To produce modulation in these devices, *p-i-n* diodes are used. They can handle up to 10 kW of pulsed power and hundreds of watts CW (continuous wave) when used as switches.

The *p-i-n* diode basically is a variable resistance. The value of resistance depends on the amount of forward bias. When the diode is used as a switch, the extremes are used; that is, heavy forward bias produces low resistance or an "on" condition, and reverse bias a high resistance or an "off" condition. When it is used as a modulator, the continuous variation in forward resistance is employed. Figure 12-3 shows the basic schematic of a *p-i-n* diode. The center portion is an intrinsic (nondoped) portion of silicon. This is therefore a high-resistance region. At low values of forward bias or at zero bias the number of carriers in the intrinsic region is small, producing low conductivity or high resistance. As the bias is increased, conductivity increases and the resistance decreases. In microwave applications, the diode looks like a finite resistance and is capable of being changed with dc. It all may seem confusing to the student who is not familiar with microwave technology, but let it be sufficient to say that the *p-i-n* diode looks like a variable resistance at high frequencies and therefore permits modulation of frequencies in the gigacycle range.

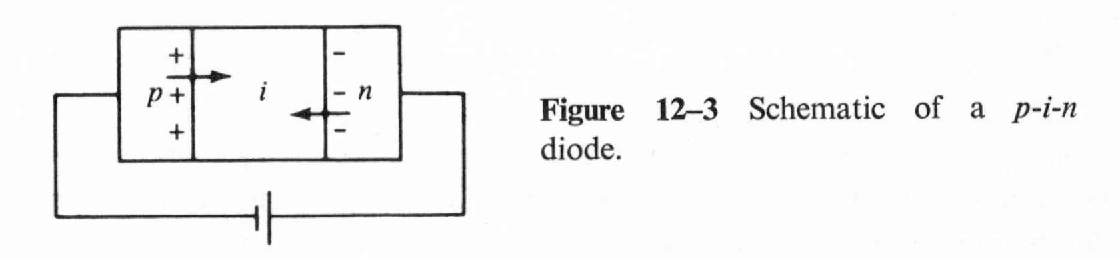

Figure 12–3 Schematic of a *p-i-n* diode.

In reverse bias a very small voltage will deplete the *i* region of carriers. In an ordinary diode the removal of carriers is not quite so quick; as a matter of fact, the amount of depletion is a function of voltage. This phenomenon of a depletion region sandwiched by carrier regions simulates a capacitor, and changing the reverse bias changes the capacitance. Since sweeping out of carriers is quite abrupt in *p-i-n* diodes, the capacitance is practically constant. Incidentally, it is this abrupt feature of *p-i-n* diodes that permits it to be used as a switch. In going from "on" to "off," only a relatively small reverse voltage is required.

The *p-i-n* diode is quite new and, at present, many applications will not appear until refinements are made. Other semiconductor devices are better known and indeed are used more extensively. One of these is the previously studied Zener diode. Another is the tunnel diode, which is covered in the next section.

12-3 The Tunnel Diode

The tunnel diode is a special type of diode used in high-frequency amplifiers and oscillator circuits. Despite its fast switching capability it is not used in digital computers, as might be expected, because other devices, such as transistors, have managed to keep pace in response time.

A typical tunnel-diode characteristic is shown in Figure 12-4. A quick glance at the curve indicates its similarity to the unijunction transistor. Figure 12-4 also shows the presence of a negative-resistance region, which in turn implies that the tunnel diode can be used as an oscillator. In addition, the tunnel-diode characteristics are similarly described; that is, we have I_P, V_P, I_V, and V_V which are peak current, peak voltage, valley current, and valley voltage, respectively.

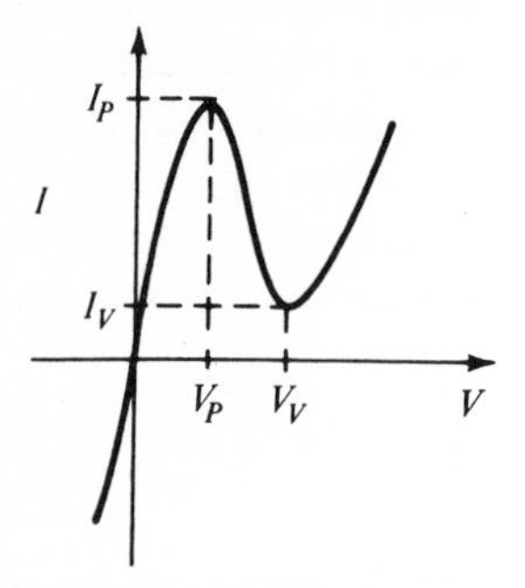

Figure 12–4 *I-V* characteristic of a tunnel diode.

The internal action, called tunneling, that causes the peculiar response curve is somewhat difficult to explain. First of all, in a tunnel diode doping is different from that in an ordinary diode. In a conventional diode one side is more heavily doped than the other, whereas in a tunnel diode both sides are heavily doped. The net effect is that conventional diodes have a wide depletion region (about 5×10^{-5} inch) and the tunnel diode has a very narrow one (about 5×10^{-7} inch). This narrow depletion region allows electrons to "tunnel" from the valence band to the conduction band or, putting it another way, the probability that an electron will go across the narrow depletion region is increased. As the voltage is increased from zero, tunneling electrons increase to a peak of I_P in Figure 12-4. This forward bias theoretically decreases the depletion region, but majority carriers (electrons in *n* material and holes in *p* material) are not released for conduction because we require approximately 0.3 volt for germanium and 0.6 volt for silicon before conduction starts. All this tunneling takes place at voltages less than the preceding two quantities. A typical peak voltage is 0.07 volt for silicon, which indeed is less than the 0.6 required for majority-carrier conduction. As the voltage is increased, the tunneling current decreases. This is the negative resistance region. Further increases in voltage results in no tunneling current. However, at this voltage the diode begins to behave like a conventional diode because we are in the range of 0.3 or 0.6 volt. At this point, majority carriers produce the necessary current. Typical values of valley-point current, where the tunnel diode begins to act like an ordinary diode, are 0.35 mA for germanium and 0.45 mA for silicon. Notice that the characteristic from the valley point on looks like an ordinary diode characteristic.

Tunneling current has an effect in the reverse direction also. Figure 12-4 shows a rather steep rise in current, whereas in conventional diodes reverse current is rather small and the response is quite flat. The sharp rise also starts from zero. These two factors permit the diode to be used as a rectifier (or detector) where signals are small primarily because there is no wait for a 0.3- or 0.6-volt rise in signal voltage before the diode decides to "produce."

Besides silicon and germanium, gallium arsenide and gallium antimonide are sometimes used in tunnel diodes. Typical high frequencies for these are 50 GHz for germanium and gallium antimonide, 1 GHz for silicon and 20 GHz for gallium arsenide. Each has its own value of peak and valley voltage as well.

12-4 The Thermistor

Generally, ordinary resistors are made with the smallest possible amount of temperature sensitivity; that is, a temperature change causes a very small change in resistance. Thermistors, on the other hand, are made to be highly sensitive to temperature change (approximately ten times more than wire-wound resistors). Materials that produce this kind of response are mixtures of pure oxides of nickel, strontium, titanium, magnesium, and cobalt.

Thermistors are manufactured in various sizes and shapes, which include disks, wafers, beads, rods, and washers. The selection depends on the application. In addition, they are supplied as a two-terminal device (directly heated) or as a four-terminal device (indirectly heated). The latter has a heater element enclosed.

Essentially, a thermistor can be used where temperature is to be sensed. In some applications the smallness or shape of the thermistor makes it the ideal device for measuring temperature. Incidentally, the word itself describes its response—namely, a resistor that depends to heat (thermo). Most thermistors are manufactured with a negative response (an increase in heat causes a decrease in resistance), although positive response is possible. Figure 12-5 shows how resistance changes in a typical case. Notice that the response is not linear and that the change in resistance is quite considerable. As an example, a temperature rise from 25°C to 100°C changes the resistance from approximately 2000 to 200 ohms, a factor of 10. Manufacturers usually quote the 25°C resistance, which is called the cold resistance.

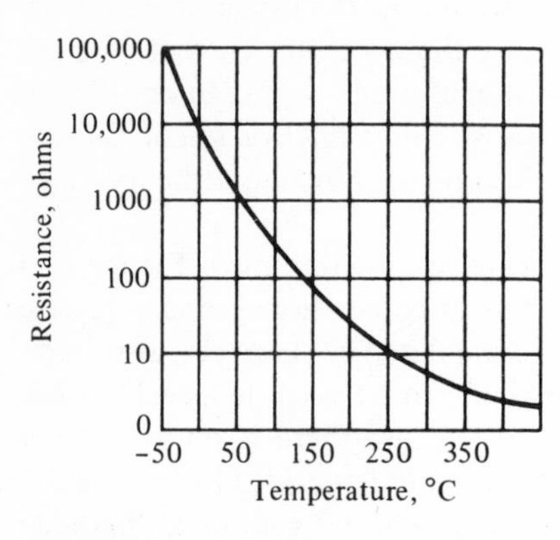

Figure 12–5 Thermistor characteristic, resistance versus temperature.

Figure 12-6 illustrates a simple temperature sensing application. In this series circuit, a temperature rise causes the thermistor resistance to drop and consequently the ammeter deflection increases. If a standard thermometer is used as a calibrator, the microammeter scale can be replaced by a temperature scale. Resistor R is used for calibrating. A more precise detection of temperature change can be realized if the thermistor is part of a Wheatstone bridge circuit. A small rise in temperature causes the bridge to become unbalanced because the thermistor resistance changes.

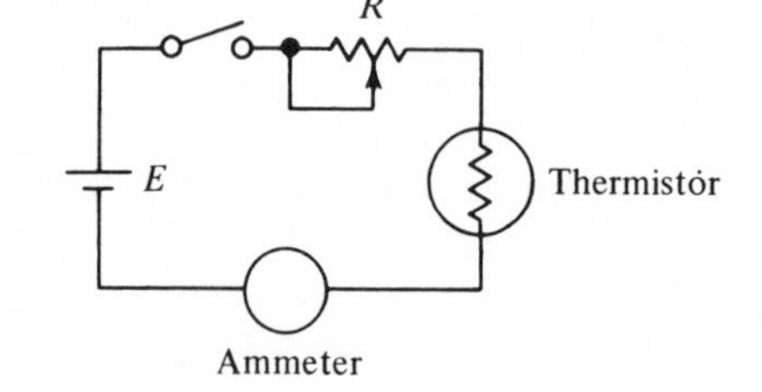

Figure 12–6 Simple circuit using a thermistor for temperature sensing.

It is not the intent of the preceding paragraph to imply that thermistors are used only as temperature-monitoring devices. Indeed, they can be used wherever a change in temperature can be nullified if that change is detrimental to circuit operation. As an example, Figure 12-7 shows how a thermistor may be used to prevent thermal runaway in transistors. First of all, the thermistor must be mounted close to the transistor so that it "feels" relatively the same temperature changes that the transistor sees. From pre-

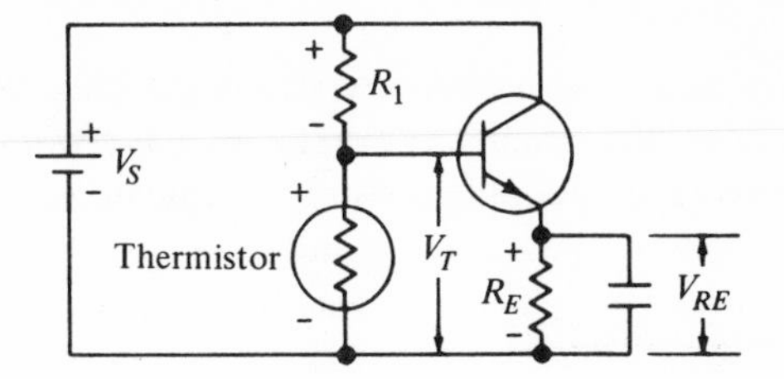

Figure 12–7 Thermistor used in biasing a transistor for temperature stability.

vious study of transistor bias we recall that the emitter–base diode "sees" the voltage V_T across the thermistor minus the voltage across R_E. V_T must be greater than V_{R_E} (by 0.3 or 0.7 volt) in the initial room-temperature design. If temperature increases, the transistor current tends to increase. However, thermistor T feels the same temperature rise and drops in resistance. Thus V_T will be less because its proportionate share of V_S is smaller. A smaller V_T means that the forward bias is decreased, and a lower forward bias results in a small transistor collector current. This offsets the impending current rise caused by temperature increase.

Another variable resistor, which seems related to the thermistor, is the varistor. It is a *voltage-sensitive* device and thus it is, of course, quite unlike the ordinary resistor, whose value does not depend on the voltage applied. The concept can be best illustrated by equations. A linear resistor follows Ohm's law,

$$I = \frac{E}{R} \tag{12-2}$$

A varistor is approximated by the equation

$$I = KE^n \tag{12-3}$$

where n is a constant which varies from 3 to 7. Representative plots of both equations are shown in Figure 12-8. As voltage across the varistor increases

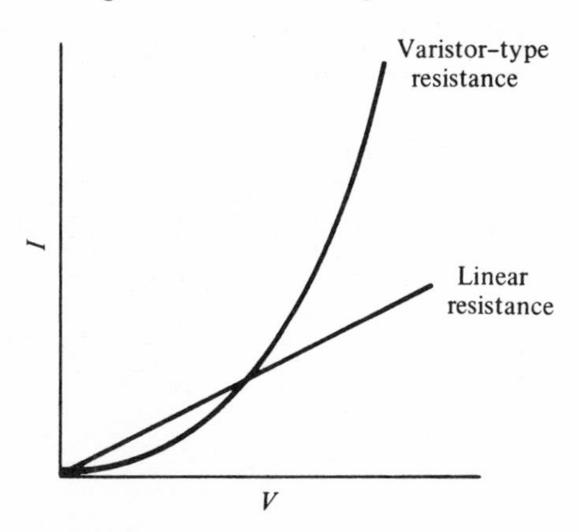

Figure 12–8 Characteristic comparison of a linear resistor and a varactor.

the resistance decreases and net current increases. This feature makes the varistor desirable in certain circuits. It may be used as a shunt across contacts to prevent arcing or as a regulating resistor to regulate voltage or current.

12-5 The Varactor

The word varactor comes from *var*iable re*actor*. In solid-state technology it is a two-terminal device, a diode whose capacitance changes as the voltage applied changes. Simply putting it, a diode is usually reverse-biased and the capacitance that results is used in various circuits. To date this phenomenon is utilized in RF tuning circuits, frequency multipliers, and high-frequency amplifiers. Before explaining how a reverse-biased diode can do the above let us first investigate how the capacitive effect is produced.

Figure 12-9(a) shows what was introduced in Chapter 2—namely, that a diode consists of a single crystal with two doped regions. In the *p* region the majority carriers are holes, whereas electrons are majority carriers in the *n* region. These majority carriers are loosely attached to the remaining portion of the atom (circled portion) and are therefore somewhat ready to move. Near the junction, there are no mobile majority carriers because in the process of formation, a combination process took place. This region is called the depletion region. In Figure 12-9(b) a reverse bias causes a larger depletion region because the mobile majority carriers are attracted away from the junction by the battery. A larger reverse bias further increases the depletion

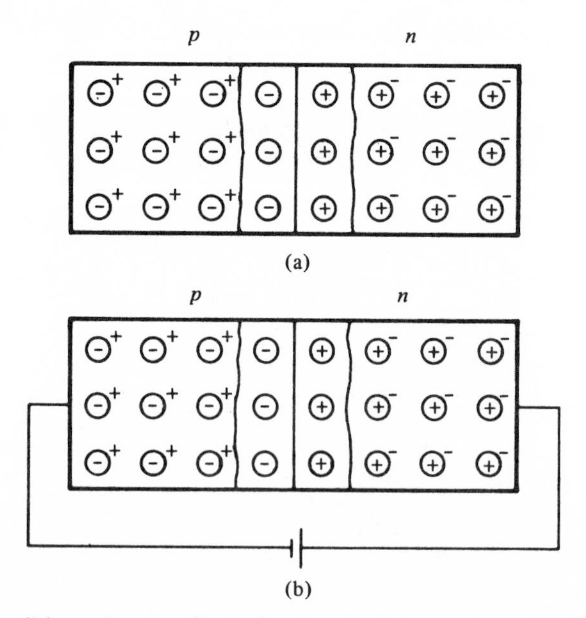

Figure 12–9 Schematic of a diode showing depletion region with (a) zero bias and (b) some reverse bias.

region. Since this depletion region has no mobile carriers, it behaves as an insulator. The carriers, in turn, are crowded nearer the ends of the semiconductor and this increased density makes these regions behave as conductors. So we have two conductors separated by a dielectric—which, incidentally, describes a capacitor.

It is well known that a capacitor can have its capacitance value changed by changing area, dielectric, or distance between plates. Equation (12-4) summarizes this relationship:

$$C = \frac{kA}{4.45d} \tag{12-4}$$

where C is in picofarads when A is in square inches and d is in inches. In a reverse-biased diode, d changes as the voltage changes; that is, if the reverse bias is increased, the depletion distance d increases and the capacitance decreases proportionately. This phenomenon occurs with most reverse-biased diodes, but only diodes specifically designed as voltage-variable capacitors are called varactors.

A typical response to voltage is shown in Figure 12-10(a). Notice that the

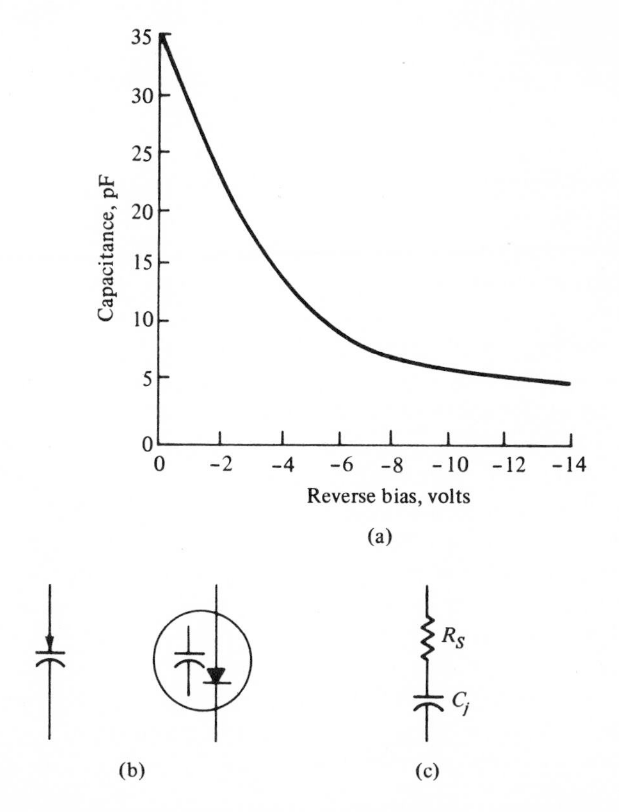

Figure 12–10 (a) Varactor capacitance versus reverse bias. (b) Varactor symbols. (c) Equivalant circuit.

response is nonlinear. This nonlinear feature is utilized, as will be explained later. Crossing the zero-volts axis (forward bias) will continue the capacitance effect for a short voltage distance. In most applications the varactor is reverse-biased only because it draws very little current and hence consumes little power. Figure 12-10(b) shows two frequently used symbols and Figure 12-10(c) shows a simplified equivalent circuit. R_S is the series resistance offered by the semiconductor material and attached leads.

Manufacturers usually quote parameters that further discriminate between varactors. For example, the Q is a quote quantity that relates how big the capacitive reactance of the diode is relative to R_S at a particular frequency. All this means is that a high-Q varactor has low resistance relative to the reactance, and therefore less power loss will occur. Since it decreases with frequency, the value of Q is quoted at a particular test frequency.

Also, manufacturers usually quote nominal capacitance. This quantity is given for a specific frequency and voltage. This value can be anywhere from 0.1 to 2000 pF. Since this changes with voltage, and a range is usually desired in most applications, manufacturers quote a capacitance ratio. This is the ratio that can be expected if the maximum-voltage capacitance is compared to the minimum-voltage capacitance.

The chief advantage of using varactors is that they simplify and improve some circuits. As an example, most of us are familiar with the air capacitor used in radios to tune in various stations. Here area change is used to change capacitance as plates are meshed or unmeshed. Physical action changes capacitance and one is relatively restricted to the location of the bulky capacitor for tuning purposes. On the other hand, a voltage-variable capacitor is simpler and can be located anywhere because a variable dc voltage can be made available almost anywhere.

Varactors are used when frequency multiplication is desired. Simply putting it, frequency multiplication is a process whereby a multiple of a particular frequency is achieved by sending the fundamental through a stage and then tuning for the multiple. As an example, a 2-MHz output is obtainable if a 1-MHz signal is introduced to the input of a varactor multiplier stage. For that matter, 3 or 4 MHz is obtainable if a tuned circuit at the output is tuned to either of these frequencies. However, the higher the multiple tuned for, the lower the expected amplitude. All of this occurs because the varactor is nonlinear. Therefore, the output will be distorted with respect to the input. Suffice it to say at this point that the varactor acts as if multiples of the input frequency are also mixed in as the input frequency passes through. Putting it another way, a distorted wave, similar to the one the varactor produces, occurs whenever a fundamental frequency has multiples of the fundamental mixed in. If indeed this is true, then it is permissible to extract these multiples by connecting a tuned circuit and tuning it to the desired multiple.

It was also mentioned earlier that the varactor serves as an amplifier.

Indeed, it may seem somewhat confusing to expect a diode to amplify. To understand this phenomenon we must look at a fundamental relationship of capacitors:

$$V = \frac{Q}{C} \tag{12-5}$$

If Q (not the Q previously mentioned, this stands for a change in coulombs) is to remain constant, a change in V must accompany any change in C. In a varactor amplifier this is precisely what is done. A signal effects a small capacitance change, which in turn produces a voltage change. A comparison of signal to varactor voltage change produces a gain. Figure 12-11(a) shows a simple varactor amplifier; Figure 12-11(b) is a simple triode amplifier and it shows some similarities to the varactor amplifier. Basically, tube T offers resistance to E_{BB}. This tube resistance is in series with R_L. When a signal appears on the grid it essentially varies the tube resistance. In the process the voltage across R_L changes at the signal rate; moreover, it is larger than the signal voltage. Hence the system amplifies. In Figure 12-11(a) no dc is used, but instead a high-frequency power supply (at least 10 × the signal frequency) is used, with R_L and varactor V_1 in series. (C_1 is chosen to have low reactance to the high frequency and high reactance to the signal. It blocks the signal from entering R_L and forces it to enter the varactor.) Varactor V_1 offers some reactance to the high frequency. When the signal enters the varactor it changes the reactance at a low-frequency rate. (L_1 blocks the

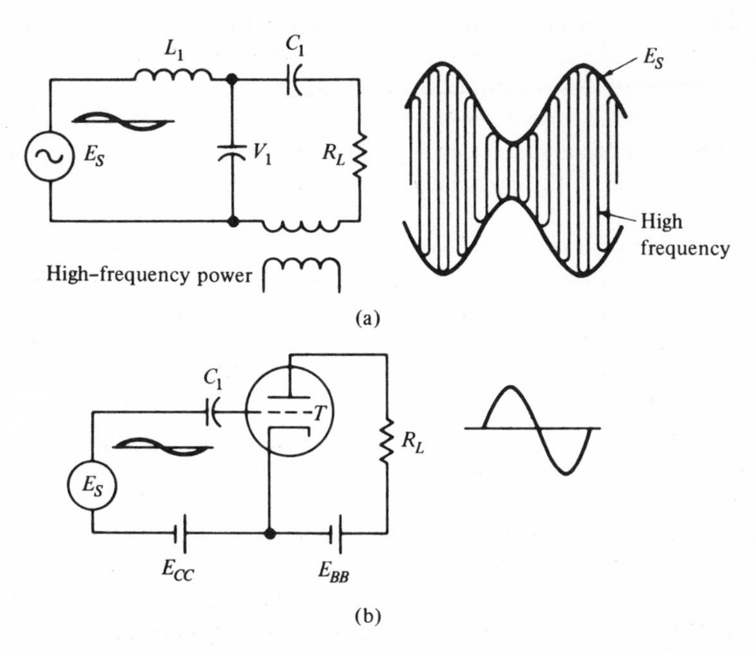

Figure 12–11 (a) Varactor amplifier. (b) Conventional triode amplifier.

high-frequency energy from entering the signal source.) High-frequency current is now modulated at a low-frequency rate, as shown in Figure 12-11(a). In the process the low-frequency amplitude at R_L is larger than the input, and therefore the system amplifies. A subsequent stage removes the high frequency from the low. L_1 is chosen so that it resonates with V_1. In most applications, a dc bias puts the operation of the resonant circuit slightly off resonance. For this reason this type of varactor amplifier is called a resonant-slope amplifier. Another frequently used varactor amplifier is a parametric amplifier, in which the varactor offers reactance to the signal and is varied at a high-frequency rate. Its principle of operation is opposite to that of the resonant-slope type. Incidentally, the parametric amplifier derives its name from the fact that a device parameter is varied to achieve amplification. In this case capacitance is the parameter that is varied to achieve the desired results.

12-6 Light-Emitting Diodes (LEDs)

Diodes that emit light when voltage is impressed behave just like incandescent lamps, but without the heat that accompanies the latter. In some applications, light-emitting diodes can replace cathode-ray tubes, pilot lamps, and readout tubes. (These possibilities plus the inherent advantages of a solid-state device predict a promising future for light-emitting diodes.)

In conventional germanium and silicon diodes electrons from the *n* material are injected into the *p* region when forward-biased. Here they recombine and the energy that is given up is dissipated in the form of heat. Other semiconductor materials, such as gallium arsenide, also produce holes and electrons when forward-biased. However, here some of the recombination energy is given off as radiant energy. These bundles of light energy, or photons, travel through the crystal and are either reabsorbed or escape from the crystal as light.

The wavelengths that are favored depend on the material. Pure gallium arsenide covers a range of wavelengths and peaks near 9000 Å. This range makes it particularly compatible with infrared detection devices. Doping gallium arsenide with certain materials will produce visible red light. Also, gallium phosphide produces green light and, with proper doping, can be used to yield red light.

Obviously, it is of some interest to see how light energy manages to escape the *p-n* junction of these materials. Keep in mind that for these special materials light energy is produced and somehow must escape from within so that it can be visible. Figure 12-12 is a simple illustration of a flat type. Light generated at the junction escapes if it follows paths 1 or 2 in Figure 12-12(a). If the bundle of light strikes the surface at an angle as shown by path 3, it is reflected back into the material and is reabsorbed. How much escapes as visible light depends on how many photons follow path 3 or, in

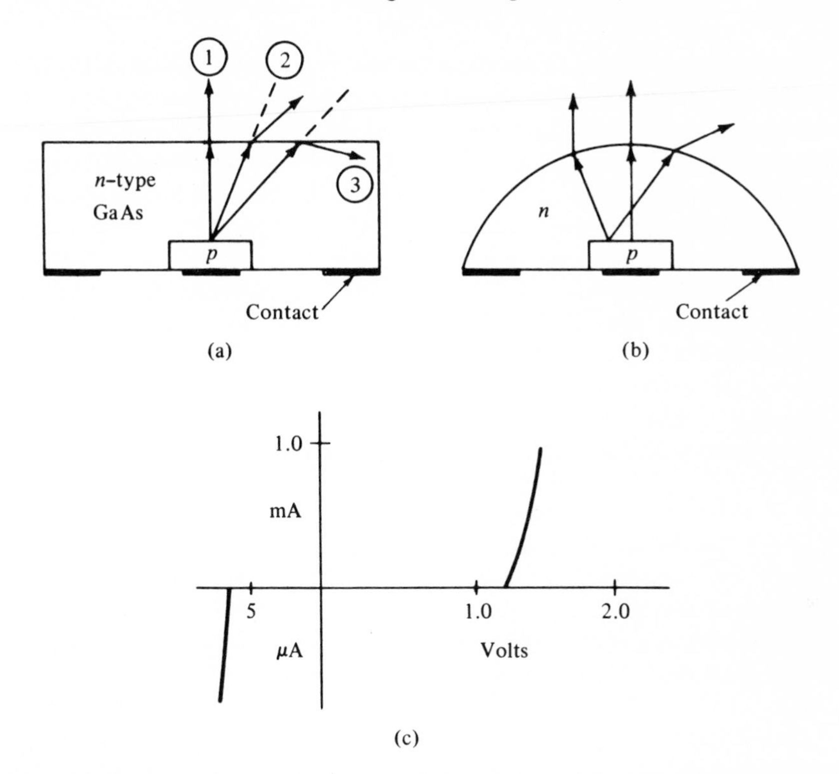

Figure 12–12 (a) Flat gallium arsenide light-emitting diode. (b) Hemispherical-shaped LED. (c) Gallium arsenide *I-V* characteristic.

other words, on the index of refraction of gallium arsenide. More reflection means less visible light. To improve the amount of emitted light, a configuration as shown in Figure 12-12(b) is used. In this case, gallium arsenide has a hemispherical shape and internal reflection is minimal. This type improves the output by a factor of approximately ten.

Figure 12-12(c) shows a typical *I-V* gallium-arsenide characteristic. Illumination takes place when the diode is forward-biased above 1.3 volts.

One strong drawback of LEDs is the cost. The unit price is relatively high because of the gallium arsenide price. However, as with other solid-state devices, further developments, refinements, and increases in demand will eventually decrease price.

In optical systems application of LEDs requires investigating the detector as well as the emitter or LED. This of course means knowing the wavelengths emitted and ensuring that the receiver or detector is sensitive to these or part of these wavelengths. In addition, an LED performance or radiant power drops off as distance from the LED is increased. So the problem of matching LEDs to detectors is somewhat complex.

One final word is necessary to show how some problems are overcome by using LEDs. Detectors are sometimes confused by ambient light in an optical system. This is to say that a detector may respond to a room light as well as to the LED to which it is designed to respond. This of course is undesirable. To overcome this drawback, LEDs are pulsed and the detector is designed to respond to pulses. The usual background lights may still be on because they do not have high-speed pulse capabilities of the LEDs.

12-7 Hall-Effect Devices

Hall-effect devices fall into the realm of semiconductors because the basic material is a thin slice of semiconductor, such as indium antimonide. The principle of operation follows with the aid of Figure 12-13. A constant current I is passed lengthwise through the material. A magnetic field B, at right angles to the current, has a peculiar effect. It tends to bend current carriers toward one edge. This accumulation of carriers produces a voltage V_h across the narrow edges. The net result is that either current or a magnetic field can be monitored by the magnitude of voltage V_h. This is more apparent by investigating a basic equation which relates all quantities. The equation is

$$V_h = KIB \sin \theta \tag{12-6}$$

where K is a constant for the material, I is the current through the material, B is the magnitude of the field (flux density), and θ is the angle between the current and the magnetic field. This latter function, $\sin \theta$, tells us that a 90-degree angle will produce a large voltage and when the field is "in line" (zero degrees) with the current no voltage will be produced. In any case, this type of device lends itself nicely to the measurement of magnetic fields. Once a Hall-effect instrument is placed in the field for maximum deflection

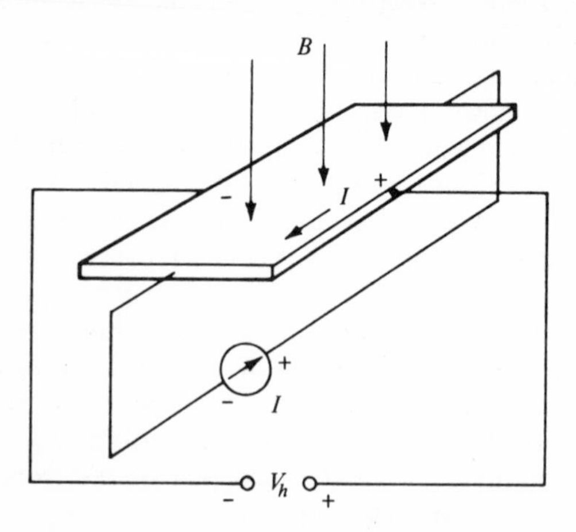

Figure 12–13 Schematic diagram of Hall-effect device.

on the voltmeter, all that is necessary is to read this deflection and convert it to a magnetic quantity. Conversion is easily accomplished by calibrating the instrument and the voltmeter in magnetic units.

It is interesting to point out that the Hall-effect phenomenon has been known since 1879. Early materials were antimony, zinc, and cobalt. The problem that plagued the usefulness was the sensitivity; that is, the output Hall voltage was relatively small. In general, all these materials gave poor response because of the mobility of the carriers within. With the advent of semiconductors, techniques were developed to yield high mobility without reducing the internal resistance of the device. Hence many areas for application were open. Some other areas, besides magnetic field monitoring, are power measurement, ammeters, and hysteresis-curve tracing.

12-8 Summary

This chapter covered some other solid-state devices. A few, like the Zener diode, are quite common in present-day circuitry; others are still being developed and are quite specialized, like the *p-i-n* diode, and are not so common. In short, Zener diodes, tunnel diodes, and varactors in all probability will be seen more frequently than the others mentioned in this chapter. In most instances, the student will first recognize the symbol in a given circuit and then will find it necessary to recall the principles that govern the functioning of these special items.

Reviewing these devices, we first looked at the Zener diode, in which the reverse-voltage breakdown region of a diode is employed for a useful purpose. The voltage remains quite constant as current through it varies. This characteristic permits the Zener to be used in regulating circuits wherein load current changes do not necessarily reflect large changes in voltage. Because of this steady voltage, other uses of the Zener diode have made it common in present-day circuits. Some of these applications are in bias circuits, clipping circuits, and surge-voltage protection.

The next device studied was the *p-i-n* diode. Its chief use is in high-frequency work, where modulation or switching is required. The *p-i-n* diode obtains its peculiar response because the *p* and *n* regions are separated by an undoped region. As a matter of fact, different responses of diodes are realized by changing doping profiles. In those profiles that result in high carrier mobility, high-frequency applications are possible. In any event the *p-i-n* diode falls into a high-frequency category and changing the forward bias changes the resistance quickly. Therefore, it is applicable for high-frequency modulation.

The tunnel diode is another device whose doping profile produces a peculiar effect. Because it is heavily doped in both regions with a narrow depletion region, a peculiar effect is evident in the *I-V* characteristic. This effect, a negative-resistance region, is produced because tunneling of carriers occurs at very low voltages. Taking advantage of this negative-resistance region permits the tunnel diode to be used in oscillator circuits. Once past the negative-resistance region, the tunnel diode behaves like a conventional diode. This implies that the region for tunnel operation is at a low voltage—for example, less than about 0.3 volt for germanium tunnel diodes and less than about 0.6 volt for silicon.

In the next section we studied thermistors. Here we realized that certain materials produced a large negative temperature characteristic; that is, as temperature increased the resistance decreased. This phenomenon permitted sensing temperature electrically. The resistance change was quite large for temperature changes; however, it was nonlinear. This required calibration with some standard thermometer before relying on the thermistor results. In addition, the fact that thermistors come in various shapes permits flexibility in applications.

Varactors were studied in the following section. The doping profile allows the use of diodes as variable capacitors. Reverse bias separates the mobile carriers. The region between the carriers is void of carriers and this reverse-bias diode simulates a capacitor. Increasing the bias increases the depletion region, which means a decrease in capacitance. The advantage of varactors are quite obvious in tuning circuits. Instead of using the typical variable-area tuning, a change of voltage can be used to yield the same effect. In addition, the varactor, which is a variable capacitor diode, lends itself to use as a parametric amplifier. Although it is somewhat difficult to imagine that a diode can amplify, the operating principle of this device involves changing a parameter to obtain amplification. In the case of the varactor, C is changed. The varactor amplifier is the most practical paramatric amplifier.

Recent years have shown a rapid development in light-emitting diodes. Special material such as gallium arsenide exhibit this phenomenon. These materials produce light when holes and electrons combine, whereas conventional diodes produce heat. In any case, the problem of permitting light to escape the crystal is efficiently solved by the use of a hemispherical rather than a flat configuration for the diode. The range of emitted wavelength varies from visible to infrared depending on the doping material. It is necessary to match the receiver or detector to the emitter in any optical system.

Finally, we investigated the Hall-effect device. In this device current goes through and an externally produced magnetic field crosses the current. The net result is a voltage that depends on the magnitude of current, magnitude of magnetic field, and the angle between the current and field. This type of device lends itself well to the measurement of magnetic-field strength.

Questions and Problems

1. Sketch a complete I-V characteristic for a Zener diode. Identify the Zener region.

2. Sketch on this curve the Zener responses for a 6-volt and a 12-volt Zener diode.
3. A shunt regulator system is used as shown in Figure 12-2. Determine a value for R_S if $I_L = 50$ mA, $V_Z = 18$ V, and $V_G = 40$ V.
4. If the load current decreases from 50 mA to 30 mA, what happens to Zener current? What is the magnitude of this change?
5. Under the conditions of Problem 3 what power is dissipated in the Zener diode?
6. Under conditions of Problem 4, what power is dissipated in the Zener diode?
7. Analyze Problem 3 and determine the maximum power rating the Zener diode must have. (*Hint:* assume all the current goes through the Zener.)
8. Why is a *p-i-n* diode so designated? Where are *p-i-n* diodes used?
9. Sketch a tunnel diode *I-V* characteristic.
10. At approximately what voltage do tunnel diodes behave like conventional diodes for germanium and silicon?
11. What is the region called that lies between the peak-voltage and valley-voltage points?
12. Why is a thermistor so called?
13. What kind of coefficient, negative or positive, do thermistors have?
14. How does this coefficient compare in magnitude to that of ordinary resistors?
15. Why is a varactor so called?
16. Explain how variable capacitance is achieved in a diode.
17. Explain how tuning is achieved in present-day radios and how a varactor can replace this conventional tuning method.
18. What advantages are gained in Problem 17 by using a varactor?
19. What is a Q of a varactor?
20. If resistance increases, what happens to Q?
21. Why is a varactor diode rich in harmonics?
22. What is a parametric amplifier?
23. How does a light-emitting diode differ from a conventional diode?
24. What is a common semiconductor material that is used in LEDs?
25. What precaution is foremost when light-emitting devices are used in an optical system?
26. Describe how a Hall-effect device works.
27. When is maximum Hall voltage developed if I and B are constant?
28. Where are Hall-effect devices primarily used?

13

Electron Tubes

13-1 Thermionic Emission

Despite the fact that all previous chapters dealt with semiconductors, vacuum tubes still play an important role in modern-day electronics. One quick example where semiconductors have not taken over is in the television picture tube. This huge vacuum tube has not been economically duplicated by solid-state devices. This does not mean it is not possible. Indeed, someday we can expect a flat screen and in all probability it will be a solid-state device.

Another area in which transistors have not taken over is in commercial transmitters. The power levels are so high that comparable performance has not yet been achieved with solid-state devices. It is therefore, necessary to study some of the characteristics of tubes and perhaps make a mental comparison between them and transistors.

To begin with, most electron tubes are thermionic devices; that is, they require heat to make electrons available. You recall that the principle of transistors was to establish a source of electrons, and then gain control over them. The same process applies in electron tubes except that we require heat to generate the electrons. Certain materials, when heated, can give up electrons much easier than others. Essentially, when a metal is heated the atoms acquire energy. Some outer-orbit electrons are so "agitated" by this increase of energy that they leave the surface of the metal. If these "boiled off" electrons escape into a vacuum, they become available to flow in a circuit. This, of course, depends on some kind of force that will attract or motivate the electrons to move once they are out of the original material.

How much energy is necessary to "excite" electrons out of its "home" metal? This depends on the material. Obviously, materials that require as little energy as possible are used. Typical metals are tungsten and thorium.

Thorium requires 3.3 and tungsten 4.5 electronvolts of energy to free the electrons. An electronvolt is the energy gained by an electron as it passes through a potential of one volt. In any case, what is important is the magnitude of this energy (or heat); the larger the number the more work must be done upon the material to free the electrons into the vacuum. It is for this reason that these energy values are called *work functions*. And since tungsten has a higher work function, more heat or higher temperatures are required. Adding 1 percent thorium to tungsten lowers the work function and hence the necessary temperature. Lower work functions are obtainable if oxides of barium or strontium are coated over nickel. Work functions of 1 to 2 are thus obtained. This type of emitting material makes portable (tube-type) radios possible because batteries can supply sufficient power to heat the material. Typical temperatures are 750°C for oxide-coated materials and 1600°C for tungsten.

To obtain these temperatures, which in turn cause the electrons to "boil" into a vacuum, current is sent through a wire that contains the electron-emitting material. This type of wire is called a filament (or cathode) and this type of heating is called direct. Figure 13-1(a) shows a directly heated filament. If the electron-emitting material is coated on a separate cylinder from the heating filament, we have an indirectly heated cathode. The heating filament is called a heater; see Figure 13-1(b). The type of cathode arrangement can be determined by looking for the construction for a particular tube in a tube manual. In either case, the manual will indicate the necessary voltage and current that will produce the high temperature required for electron emission. This voltage and current is so determined that the amount of electrons emitted is more than enough for the recommended application. By this is meant that so many electrons are boiled off that a cloud of electrons exists around the cathode. This cloud, which serves as the source in typical uses, is called a space charge.

The next question that might arise is: How do we move the electrons? The answer is that since the electrons are negative, we must place within the vacuum tube a negative-seeking electrode that will attract them. If a plate,

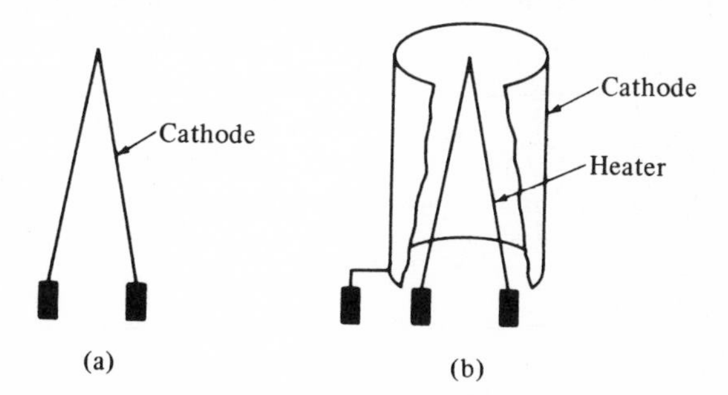

Figure 13–1 (a) Directly heated and (b) indirectly heated cathode.

or anode, has a positive potential *with respect* to the cathode, the electrons will be attracted to the anode. A simple circuit is shown in Figure 13-2(a). Battery A is in the circuit to heat the cathode. Battery B makes the anode positive with respect to the cathode. Once the electron is freed into the space charge it feels an attracting force that pulls it to the anode. Intuitively, one can see that the greater the B voltage the greater is the number of electrons attracted to the anode. This is registered on the milliammeter as a current increase. It is also conceivable that the B supply may be raised to a point where all the space-charge electrons are gone and every electron that is boiled off immediately goes to the anode. In a sense the anode is waiting for all possible electrons as soon as they are produced. In most cases, tube manual specifications limit this maximum anode voltage to a value where a space charge will always exist. Figure 13-2(b) shows a typical diode curve; the flat portion (high V_B) is the region where no space charge exists and the current depends on the recently emitted electrons.

Below this region the emission can be mathematically approximated. The equation is called Child's three-halves power law:

$$I_B = KE^{3/2} \tag{13-1}$$

where K is a tube constant that depends on anode area and distance between anode and cathode, E is the voltage applied, and I is the resulting current. A word of caution about this voltage. It is frequently called anode or plate voltage but one must remember that a voltage is always applied between two points. Figure 13-2(a) shows that it is applied between anode and cathode. So despite the fact that the word cathode is omitted, it happens to be the other voltage terminal. This is similar to transistors, where we used the term "collector voltage" where indeed we meant collector-to-emitter voltage. In any case, the circuit has to be complete if we wish to establish current flow.

A final word is in order relative to these heater or filament voltages. The first number that is encountered for a tube description relates the amount of necessary filament voltage. As examples: a 6AF4 requires 6.3 volts on the heaters; a 12AU7 requires 12.6 volts; a 50C5 requires 50 volts; and a 5Y3 requires 5 volts on the directly heated cathode. The pins in the tube base to

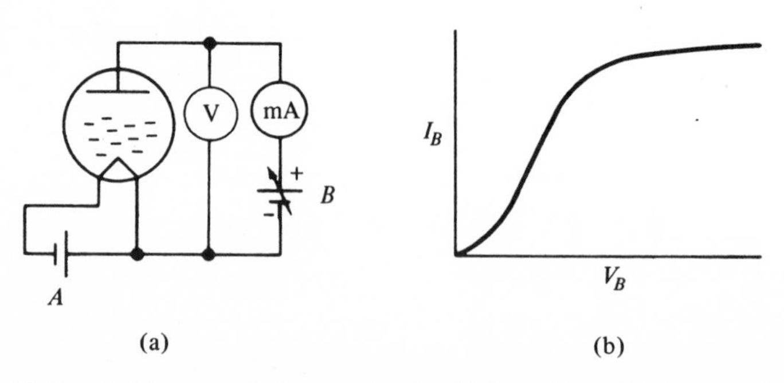

Figure 13–2 (a) Vacuum diode test circuit. (b) Resulting *I-V* characteristic.

which the heaters and cathode are connected are shown in figure form in tube manuals. Some tube bases have four pins; others may have seven, eight, or nine.

13-2 The Vacuum Diode

The preceding section described the essentials of a vacuum diode. Basically, electrons are boiled off a cathode and reach the anode in a substantial quantity when a positive terminal of a battery is connected to the anode and the negative terminal to the cathode. If the voltage is reversed, the negative electrons will see a negative anode and will be repelled; hence no current flow will occur either through the tube or externally. We experienced this behavior before when semiconductor diodes were discussed. This feature made rectification possible. Figure 13-3(a) shows Figure 13-2(a) repeated, except that in this case ac is applied between anode and cathode. This means that the anode will be alternately positive and negative with respect to the cathode. Note also that the magnitude of voltage will be changing. One can conclude that current will flow when the anode is positive and not flow when it is negative. The result is a voltage across R_L only half the time. The diode seems to act like a switch; at one instant there is a current flow and at the next instant there is not. (For this reason the British call tubes "valves.") Notice in Figure 13-3 that the ac source has a polarity. This polarity is the conduction polarity because it makes the anode positive.

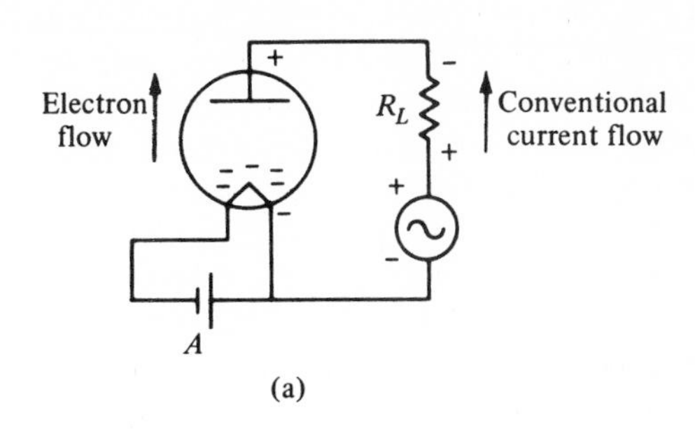

(a)

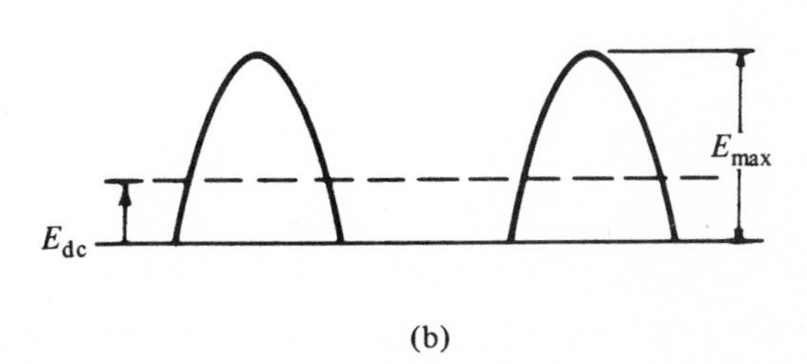

(b)

Figure 13–3 (a) Vacuum diode rectifier circuit. (b) Resulting output from a sinusoidal input.

If we stick to conventional current flow, which is current leaving the positive terminal of a source, it will be opposite direction to electron flow in the tube because we know electrons leave the hot cathode and arrive at the anode. Either approach is used in tube texts, depending on the author's preference. In either case a voltage drop occurs across R_L. This voltage is the source voltage less the tube voltage. We just simply envision the tube as a resistance (not necessarily a fixed quantity) in series with another resistor R_L, and a source that supplies the current for this voltage drop. Here vacuum diodes and semiconductor diodes differ. For the same input voltage, the voltage across R_L with a semiconductor diode will be greater than with a vacuum diode. This means that the resistance the tube offers is greater than the semiconductor. Since it is desired to get as large a voltage out as possible, the vacuum diode apparently is less efficient than the semiconductor diode. Putting it in numbers, a tube dc voltage drop may be 50 volts and the semiconductor diode only 1.0 volt. This means more of the input will appear across the semiconductor diode's load resistance.

We might add another difference between the two diodes. The vacuum diode requires heaters or a filament. These eventually become weak when the emitting material is gone; moreover, these elements sometimes burn open. These possibilities cut down the life expectancy of vacuum diodes. In addition, the heat that is generated must be dissipated sufficiently to avoid overheating. Semiconductor diodes do not have these problems to the degree that tubes have, and therefore they last longer.

Figure 13-3(a) shows a battery supplying the filament voltage. This is not usually done except in portable tube-type sets. Windings on transformers usually supply the necessary filament voltage along with the high voltage that is to be rectified. The net result is an output as shown in Figure 13-3(b). The average height of the "hills" represents the dc voltage that is produced. Obviously, if the "hills" are higher, the average or dc voltage is greater. The relationship for half-wave rectification is

$$E_{dc} \simeq 0.318E_{max} \qquad (13\text{-}2)$$

where E_{max} is the peak of the ac voltage being rectified, and we assume that the voltage drop across the tube is relatively small. This latter assumption makes the mathematical equation rather approximate because the tube drop is usually significant when low ac voltages are being rectified or R_L is relatively small.

13-3 The Triode and Its Characteristic Curves

Once the diode was understood and applied, innovations were tried that produced other types of tubes. A natural development of the diode (two elements) was the triode (three elements). The addition was a grid wire between the cathode and anode. It was placed closer to the cathode and was wound in such manner that there was plenty of space for electrons to travel to the anode. Figure 13-4 shows a sketch and a symbol of a triode.

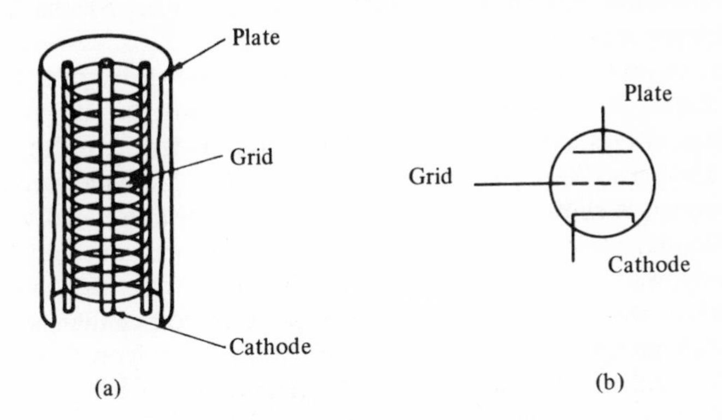

Figure 13–4 (a) Triode construction. (b) Triode symbol.

In a triode, electrons leave the cathode and head toward a positive anode. In traveling, they pass by the grid wire. This grid is usually made negative with respect to the cathode by a small dc voltage. Hence, the electrons "see" a negative grid, which is quite close, and a positive anode, which is farther away, as they leave the cathode. One element tends to repel them. Hence the amount of current that flows depends on the relative magnitudes of these voltages. In a typical setup the voltages are such that some electrons are permitted to get through; however, a slight change in the negative grid voltage controls the net flow. For this reason the grid is called a control grid. In addition, a small grid voltage change results in a relatively large anode current change, primarily because the control grid is so much closer to the cathode. This action is somewhat analogous to that of the transistor, in which a small change in base current effects a large change in collector current. And, as in transistors, in tubes we require dc potentials to set the system for operation, so that ac conditions can be satisfied. The basic circuit and dc voltages are shown in Figure 13-5. Notice that E_{BB} makes the anode positive with

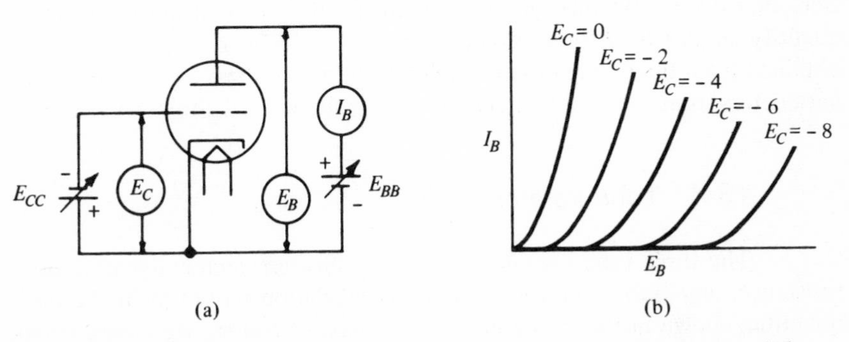

Figure 13–5 (a) Triode test circuit. (b) Typical static $I_B - E_B$ characteristics.

respect to the cathode, and E_{CC} makes the grid negative with respect to the cathode. Despite the fact that we might be jumping ahead, it seems obvious that we have just described a *common-cathode* amplifier. All that remains is a signal input arrangement and a load resistor.

But let us go back awhile and reestablish our knowledge of triodes under dc conditions by viewing characteristic curves. Just as we concluded in transistors, curves are worth a thousand words. If we set up the circuit of Figure 13-5(a) we can obtain these curves. It must be understood that the heater, which is shown but not connected, must have its rated voltage and current to make this circuit work. In any event, let us assume that E_C is zero. Essentially, we have a diode and if E_{BB} were raised from zero to some positive voltage, we would obtain a typical diode response; that is, I_B would rise as E_{BB} increased. See Figure 13-5(b) for $E_C = 0$. Now if we set $E_C = -2$ volts, grid negative with respect to cathode, what can we anticipate? First of all, the electrons are available near the cathode in a space charge. However, the influence of this anode is somewhat reduced because the grid is negative. Electrons will arrive at the anode, except that the anode will have to "work harder"; that is, now the anode voltages will have to be higher than when E_C was equal to zero to obtain the same amount of current I_B. This is seen in Figure 13-5(b), where $E_C = -2$ volts.

The process is repeated for $E_C = -4$, -6, and so on. Eventually the control grid becomes so strongly negative that any electron leaving the cathode surface is immediately repelled, and thus no electrons arrive at the plate. The total characteristic has the same relationship as transistors; that is, the coordinates are output quantities and the input quantity is in the field of the curves. One can envision a small signal voltage injected at the grid which will affect the electron stream and those electrons producing a commensurate changing voltage across an R_L in the anode circuit. A comparison of these two voltages yields the voltage gain. Not only is this arrangement similar to a CE amplifier but it is quite similar to an FET amplifier because both output and input are voltage quantities. Hence, we can conclude that the triode can be used as an amplifier and that it is voltage-controlled. Carrying the FET and transistor anology further requires obtaining parameters that distinguish one type from another. By this is meant extracting a quantity similar to β for transistors or g_m for FETs. If you recall, β was obtained from the family of curves. Similarly, with tubes, we can obtain the parameters from the family of curves. This is shown in the following section.

13-4 Tube Parameters

The three tube parameters are amplification factor, dynamic plate resistance, and transconductance. Each is a relationship between the three quantities shown in the family of curves. These, of course, are plate current, plate voltage, and grid voltage.

Amplification factor, by its phrase, implies a theoretical voltage gain. Indeed, the voltage gain is an actual comparison between output voltage and input voltage, whereas the amplification factor represents a voltage gain that is never achieved. Therefore, the amplification factor must be a ratio of the change in plate voltage to the change in grid voltage—with the extra condition that the plate current is held constant. Mathematically,

$$\mu = \left.\frac{\Delta E_B}{\Delta E_C}\right|_{I_B=\text{constant}} \tag{13-3}$$

where μ is the amplification factor. Notice the similarity of this equation with β in a transistor, where

$$\beta = \left.\frac{\Delta I_C}{\Delta I_B}\right|_{V_{CE}=\text{constant}}$$

In Equation (13-3), ΔE_B is the output quantity and ΔE_C is the input. Another meaningful way to define amplification is the ratio of a small change in plate voltage to the change in grid voltage required to restore the plate current to the value it had before the plate voltage was changed. One can conclude that the larger the amplification factor the larger the expected gain.

In Figure 13-6 a family of curves for a 6J5 triode is shown. Near point A is where the amplification factor is extracted. I_B is constant at 8 mA, and an arbitrary change in E_C of 4 volts is chosen. The resulting change in E_B is

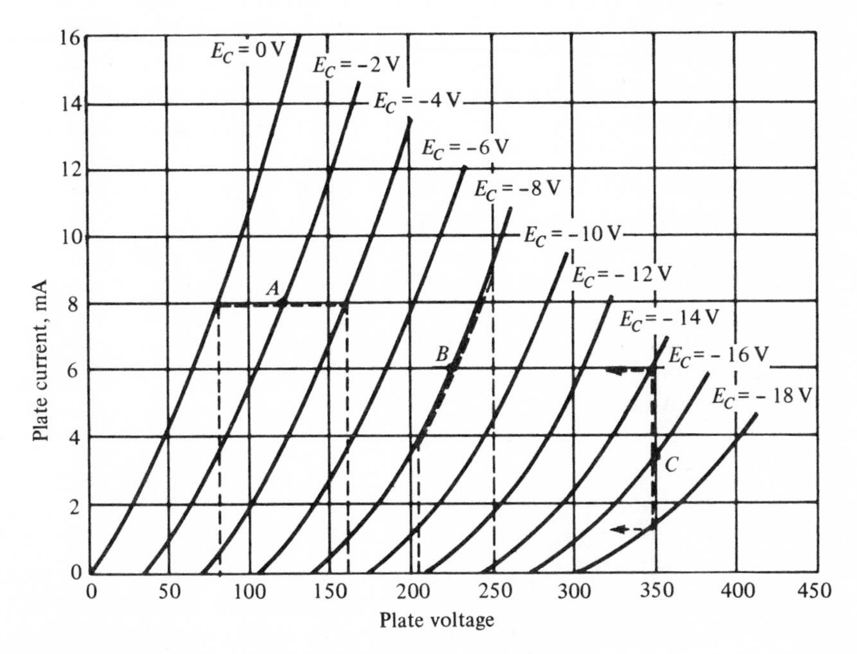

Figure 13–6 Static characteristics for a 6J5 triode.

162 − 80, or 82 volts. Therefore, solving for μ,

$$\mu = \left.\frac{\Delta E_B}{\Delta E_C}\right|_{I_B=\text{constant}}$$

$$= \frac{82}{4} \simeq 21$$

The typical amplification factor given by the manufacturer is 20 and, like β in transistors, it will be different at other locations.

The second parameter is ac plate resistance. These words imply that the ratio must be voltage divided by current to obtain resistance and it must be the plate voltage and current. Hence, these quantities are E_B and I_B and the remaining quantity must be constant which in this case is E_C. Mathematically,

$$r_p = \left.\frac{\Delta E_B}{\Delta I_B}\right|_{E_C=\text{constant}} \tag{13-4}$$

This quantity represents the amount of resistance the triode offers to ac. It too changes with location on the family of curves.

To extract r_p in Figure 13-6, point B is used. An arbitrary ΔE_B of 250 − 203 volts produced a ΔI_B of 8.9 − 4.0 mA along a constant E_C of 8 volts. Using Equation (13-4),

$$r_p = \left.\frac{\Delta E_B}{\Delta I_B}\right|_{E_C=\text{constant}}$$

$$= \frac{47}{4.9(10^{-3})} = 9.6 \text{ k}\Omega$$

A typical value quoted by the manufacturer is 7.7 kΩ.

The final characteristic necessary to describe vacuum tubes is transconductance. This parameter was used in FETs. It is similarly extracted and, as the phrase implies, it is a conductance ratio, which is I/V, and since it is "trans" or across (the tube) it must be I_B divided by E_C with the third quantity, E_B, constant. Mathematically,

$$g_m = \left.\frac{\Delta I_B}{\Delta E_C}\right|_{E_B=\text{constant}} \tag{13-5}$$

Point C is chosen to obtain these quantities. A ΔE_C of 4 volts is arbitrarily used, which produces a ΔI_B of 6 − 1.5 mA at a constant E_B of 350 volts. Substituting into Equation (13-5),

$$g_m = \left.\frac{\Delta I_B}{\Delta E_C}\right|_{E_B=\text{constant}}$$

$$= \frac{4.5(10^{-3})}{4}$$

$$= 1.125(10^{-3}) = 1125\ \mu\text{mho}$$

If points A or B were chosen to extract g_m, larger values would be realized. Point C is not a typical operating point. The given value in a tube manual is 2600 μmho. The following relationship exists between all three parameters:

$$\mu = g_m r_p \tag{13-6}$$

Thus if any two of these quantities are known, the third can be obtained; however, the two known quantities obviously must be extracted at the same place on the curve. As an example, we cannot use the μ of point A and r_p of point B and mathematically predict g_m of point C by the use of Equation (13-6).

Finally, before we leave triode parameters it is necessary to show why we bother obtaining them. If the student recalls the transistor chapter it will be obvious. In any case, there are two basic equations for the gain in a common-cathode amplifier. If the triode is treated as a constant-voltage source, the equation is

$$A_v = \frac{\mu R_L}{r_p + R_L} \tag{13-7}$$

If the triode is treated as a constant-current source, the voltage gain is

$$A_v = g_m Z_L \tag{13-8}$$

where

$$Z_L = \frac{r_p R_L}{r_p + R_L}$$

This equation of course was used in FETs. Notice that in both equations tube parameters are necessary; in addition, the same answer will be realized.

One quick look at Equation (13-7) will substantiate a remark made earlier in this chapter. Notice that $R_L/(r_p + R_L)$ is always less than 1. This fraction is multiplied by μ to obtain the voltage gain. Therefore, the actual gain of a triode amplifier will always be some quantity less than μ, and it depends on the relative values of r_p and R_L. Hence the amplification factor is some theoretical gain, which is never realized in a practical circuit.

13-5 Interelectrode Capacitance and Its Effect

One feature that was not discussed in any great detail in connection with transistors is the capacitive effect on amplifiers. This effect is much more visible in tubes and will be discussed in this section. However, in general, this phenomenon has a similar effect in transistors.

Within a triode vacuum tube, there are three active elements separated from each other by a small space. Since the elements are metallic and the space a vacuum, in essence we have a small capacitor. Any time two conductors are separated by a dielectric we have this effect. Obviously, the object of the triode is to permit amplification, but a by-product of the process is this capacitive effect. Figure 13-7 shows these capacitances. They are the grid-to-cathode, grid-to-plate, and plate-to-cathode capacitances—abbrevi-

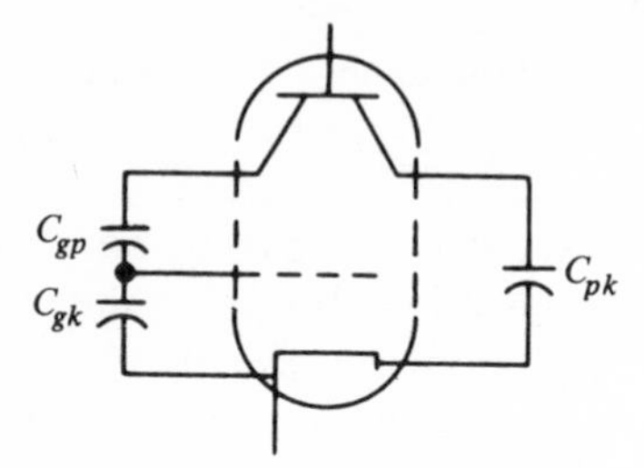

Figure 13–7 Interelectrode capacitance in a triode.

ated as C_{gp}, C_{gp}, and C_{pk}, respectively. These capacitances have a negative effect on the amplification of high frequencies. Putting it another way, an amplifier increases the amplitude of a small signal but, because of the interelectrode capacitance, the higher frequencies may not necessarily be amplified. At what high frequency this loss occurs depends on the amount of interelectrode capacitance. This quantity of course depends on the tube, and within the tube it depends on the cross-sectional area of the active elements and the distance between them. These are the basic constants for a capacitor whether in a tube or as a discrete component. Mathematically,

$$C = \frac{KA}{d} \tag{13-9}$$

where K is a constant for the dielectric, A is the cross-sectional plate area, and d is the distance between the plates. Although these values are important to the tube designer, we are more interested in the magnitude of capacitance and the effect.

Since this may be the first exposure for the student to this capacitance phenomenon, it is worthwhile to analyze the effect at a relatively simple level so that it will make sense and be easily assimilated. First of all, we must assume that this capacitance is lumped somehow into one capacitance. It is not done by simply adding C_{gp}, C_{gk}, and C_{pk}; however, we can lump them into one effective capactitance. Second, a typical tube amplifier requires a resistance between the grid and cathode. This is very similar to the gate resistance required in an FET amplifier. This resistance (R_C) and the remaining portions of a typical amplifier are shown in Figure 13-8(a). R_K is used to obtain self-bias; that is, we eliminate battery E_{CC}. R_L, of course, is the load resistance and E_S is the small signal to be amplified. As far as E_S is concerned, its two terminals do not know there is a tube "down the line" but instead it sees some kind of equivalent circuit. The equivalency of the input to the tube is shown in Figure 13-8(b) where R_{GK} is the tube equivalent resistance between grid and cathode. Since a few electrons go astray and impinge upon the grid wires instead of passing to the plate, some current does flow in the grid–cathode circuit. This equivalent resistance is 10 MΩ or more. C_L, on the other hand, represents the lumped capacitance of the triode that was previously discussed. R_S is the signal source internal resistance, which could be a signal generator or a preceding amplifier stage.

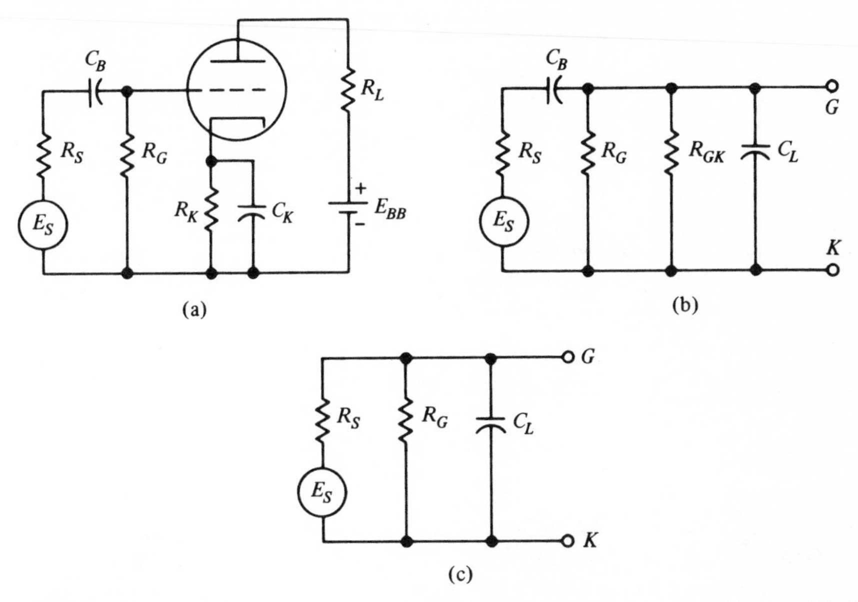

Figure 13–8 (a) Typical triode amplifier. (b) Equivalent circuit. (c) Simplified equivalent circuit.

We must reflect for a moment to keep the amplifier—with the aid of this equivalent circuit—in the proper perspective. The object of the amplifier is to increase the amplitude of a small signal. There is also a demand on the associated circuit, which is *to pass as much of the original signal as possible to the grid–cathode terminals* where the tube can do its job of amplification. Hence, if we look at Figure 13-8(b), C_B can be chosen so that it has low reactance for a low frequency and at any higher frequency that passes through the reactance will be even lower because reactance goes down as frequency goes up. This means that all of E_S will appear across R_G, R_{GK}, C_L in parallel with R_S in series. In most applications R_{GK} is much greater than R_G, so the equivalent parallel resistance is primarily determined by R_G. So if R_{GK} and C_B are dropped, the more simplified equivalent circuit will look like Figure 13-8(c). Now the amount of voltage that gets transferred from E_S to the grid and cathode terminals depends on the relative magnitude of R_S with respect to R_G and X_{C_L} in parallel. This is a voltage divider situation, and if R_G and X_{C_L} are quite large with respect to R_S then most of E_S will indeed appear between the grid and cathode terminals.

Perhaps the preceding discussion can be understood more easily if we simplify further by replacing C_L with a resistance and introduce some numbers; see Figure 13-9(a). A typical value for R_G is 100 kΩ and R_S = 50 kΩ. A quick solution tells us that the parallel combination of R_G and R_E yields a net resistance of 50 kΩ. This is shown in Figure 13-9(b). Therefore,

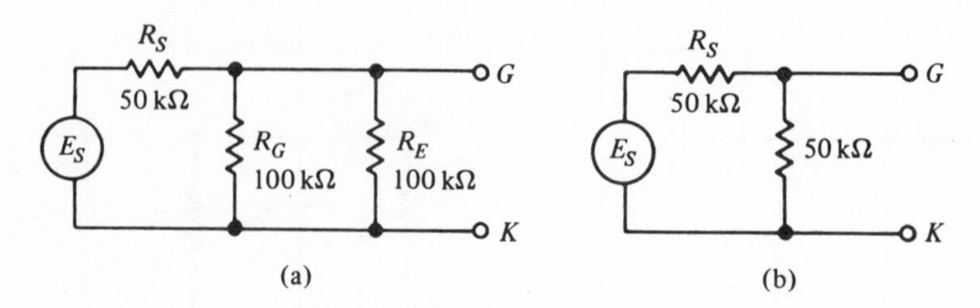

Figure 13–9 (a) Simplified circuit showing input to a vacuum tube. (b) Equivalent circuit.

the amount of voltage between the grid and cathode is 50 percent of E_S. Mathematically, the voltage division equation says,

$$E_{GK} = E_S \frac{R_p}{R_p + R_S} \qquad (13\text{-}10)$$

where R_p is the parallel combination of R_G and R_E.

Now suppose R_E is 1.0 kΩ. The net parallel resistance is approximately 1.0 kΩ and, using Equation (13-10), E_{GK} becomes

$$E_{GK} = E_S \frac{1.0}{1.0 + 5.0}$$

$$\simeq E_S \frac{1}{50}$$

This means that the voltage transferred from E_S to the G-K terminals is considerably smaller when R_E is 1.0 kΩ than it is when R_E is 100 kΩ. Putting it another way, the combination of R_G and R_E must remain high if maximum voltage is desired between grid and cathode.

Returning to a vacuum-tube amplifier, there is no R_E, but rather a C_L, as shown in Figure 13-8(c). Since the reactance of C_L decreases as frequency goes up, it is conceivable that at a particular frequency X_{CL} is 100 kΩ. However, if the frequency increases sufficiently, X_{CL} becomes 1.0 kΩ as in the foregoing simplification. Therefore, we can conclude that triode amplifiers will not amplify high frequencies because of the input capacitance effect. At what frequency this drop-off in gain starts (it is a gradual, not abrupt, decrease) depends on all the factors previously discussed. These are: R_S; the chosen R_G; and C_L, which depends on the tube makeup.

Amplifier stages employing triodes have poor high-frequency response unless a special effort is made to improve this situation. One easy way to aid high-frequency response without too much effort is to use tubes that are not triodes and that have a relatively low lumped capacitance. It is conceivable that if C_L decreases X_C will increase and thus create a more favorable parallel combination of X_C and R. The next section will investigate one tube in this category.

13-6 The Pentode and Its Characteristic Curves

The difficulties that triodes produced at high frequencies led to development of other type tubes. It seems logical that someone would place another grid wire between the control grid and the plate to reduce the interelectrode capacitance. Indeed, this was done; the capacitance was reduced with this addition of a fourth active element, the screen grid. This type was called a tetrode. To guarantee that the emitted electrons arrived at the plate this screen grid had a positive potential placed on it with respect to cathode. The potential used is always less than the plate voltage. However, one drawback resulted; that is, the *I-V* plate characteristic has a dip in it. See Figure 13-10(a). To be effective as an amplifier the tube must avoid this

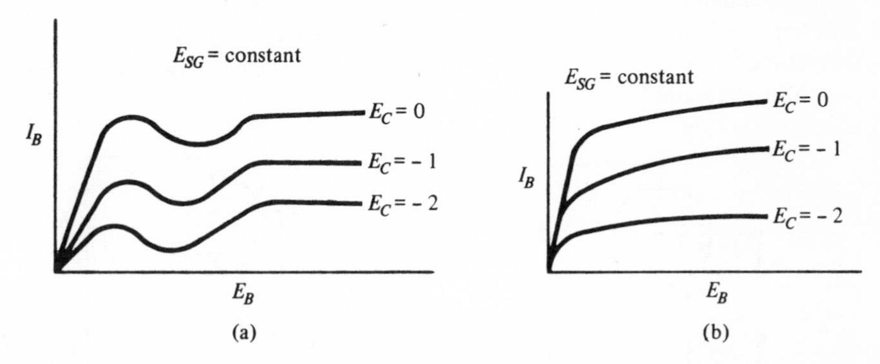

Figure 13–10. (a) Typical characteristics of a tetrode. (b) Typical characteristics of a pentode.

region. Therefore, the range of usefulness is limited. It now seems logical that again someone would introduce another grid. A five-element tube called a pentode has a grid wire between the screen grid and the plate. This suppressor grid, which is operated at cathode potential, removed the tetrode dip, as shown in Figure 13-10(b). The extended usefulness of tube operation and further reduction of grid–plate capacitance made possible by the pentode has caused the tetrode to become nearly obsolete. If we digress for a moment, notice that the tetrode has a negative slope. We first saw this phenomenon and also made use of it when unijunction transistors were first discussed. Hence, we can conclude that the tetrode can be used similarly if we operate in the negative resistance region. In any case, Table 13-1 shows typical character-

Table 13-1

	Triode	*Tetrode*	*Pentode*
C_{gp}	3.0 pF	0.007 pF	0.004 pF
r_p	40 kΩ	400 kΩ	700 kΩ
μ	60	400	1000

istics of the three tubes discussed. Notice that pentodes generally have high amplification factors and thus we can expect higher voltage gains than we can with triodes. Table 13-1 also shows a considerable decrease in grid–plate capacitance, which has a large contribution to the lumped capacitance.

So, as a quick review we can say that triodes, tetrodes, and pentodes are used in amplifier stages. We can expect higher gains from pentodes because they have higher values of μ. We can also expect higher frequencies to pass through pentodes easier than through triodes because the interelectrode capacitance is decreased. For amplifier operation, the necessary dc voltages must be placed between the proper elements. Figure 13-11 shows all the necessary potentials, along with load resistances for triode and pentode amplifiers. Essentially, the signal appears between the control grid and cathode and the amplified output between the plate and cathode. We can conclude that these are common-cathode amplifiers. Recalling transistor amplifiers, we can expect such other configurations as common-grid and common-plate amplifiers.

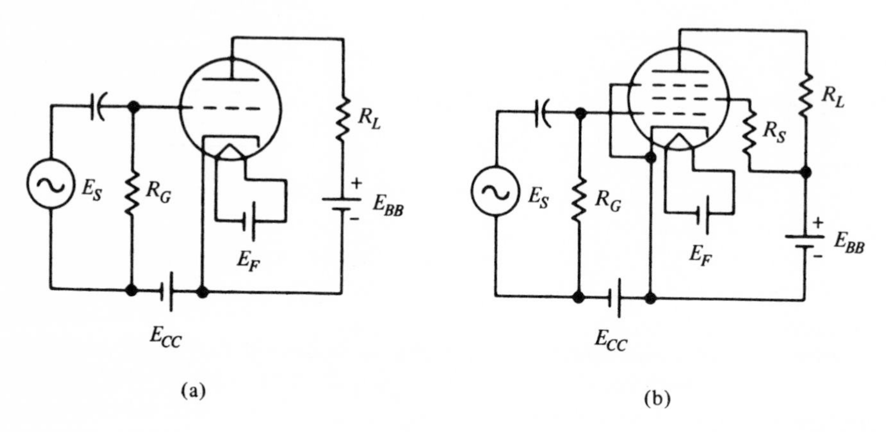

Figure 13–11 (a) Typical triode voltage amplifier. (b) Pentode voltage amplifier.

13-7 Beam Power Tubes

In many electric systems, after signals are amplified, the signals must do some useful "work." An example of this is the ordinary radio. The purpose of most of the stages is to strengthen the useful signal. Eventually, this improved or beefed-up signal must drive a loudspeaker, whose job is to convert electric power to audio power. In other words, the loudspeaker must vibrate at the signal rate. This will make the local air particles move and these air particles will vibrate our audio-sensing mechanism. Considerable power must be delivered to the speaker in order for the necessary vibrations to occur. Tubes specifically designed to deliver more power than the average tube are called power amplifiers. This simply means that the vertical axis of an *I-V* plate characteristic may have a current range of 0–400 mA, as compared with 0–10 mA for a nonpower amplifier. It also means that this tube is one of the hottest tubes in a given piece of electronic gear.

The question that arises is how is this achieved structurally? The name "*beam* power tube" is the clue. In this type of tube, the control grid and screen grid are aligned so that the electron flow from cathode to plate is in sheets or beams. Also, on the sides are beam-directing plates, which are tied to the cathode and help direct the electrons into beams in the desired region. The result of the whole construction is pentode-type response, but with very high plate current. A typical power tube is a 50C5. The student may find it interesting to look up the 50C5 in a tube manual and investigate its power rating and maximum current capability.

13-8 Variable-Mu Tubes

A variable-mu tube is a special tube that, as the title implies, has a variable amplification factor (depending on the amount of bias on the control grid). Indeed, every tube has an amplification factor that changes somewhat with operating point. However, in a variable-mu tube the change is intentional and more predictable.

Let us look at a conventional tube and plot what is called a transfer characteristic. (The word transfer must imply what coordinates are necessary for a plot.) This characteristic shows how plate current I_B decreases as the control-grid bias is increased for constant plate voltage and screen-grid voltage. Figure 13-12(a) shows that no current flows once the grid is 6 volts (or more) negative with respect to the cathode. In a variable-mu tube the cutoff is not as definite, as shown in Figure 13–12(b), and essentially the grid must go quite negative before cutoff (zero plate current) is reached. Incidentally, this tube is sometimes referred to as a *remote-cutoff* tube. If we took

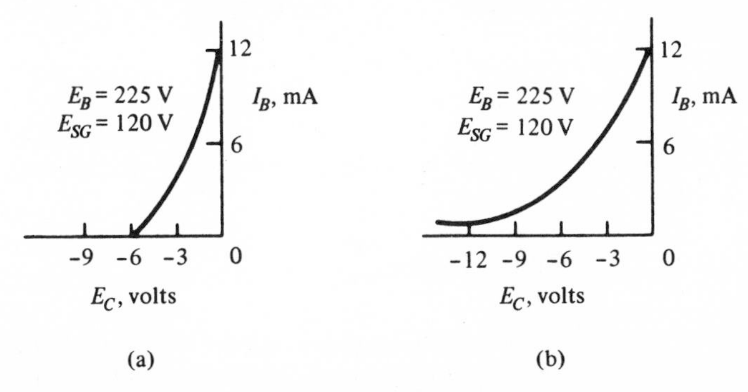

Figure 13–12 Transfer curve for (a) ordinary tube and (b) variable-mu tube.

the slope of the curve it would yield g_m. By inspection, the slope is not constant but is large near $E_{CC} = 0$ and small as we approach large E_{CC} values. Keep in mind that a slope is defined as how much the vertical changes per unit change in the horizontal direction. Hence, g_m is variable and, since μ is proportional to g_m, we can consider this type as a variable-mu tube.

The call for this type of tube is not extensive. It is used chiefly in radios where the gain can be controlled by the amount of bias because a change in bias produces a change in gain.

Once again, we can ask, how is this remote-cutoff characteristic achieved? Inside the tube, the grid wires are normally wound with a fixed number of turns per inch for the sharp-cutoff tube. See Figure 13-13(a). The net effect is that a given bias repels some electrons all along the grid. In a variable-mu or remote-cutoff tube, the grid wire is wound relatively tightly near the ends, but with fewer turns per inch near the center. See Figure 13-13(b). The net result is that the negative grid will have more effect at the ends than at the center and when sufficient bias is present no electrons will get through at the ends but, because of the spacing, a few will still manage to get through near the center. As a matter of fact, it will take a substantial negative voltage before cutoff is reached at the center. Hence, the variable pitch-wound control grid produces a variable-mu characteristic.

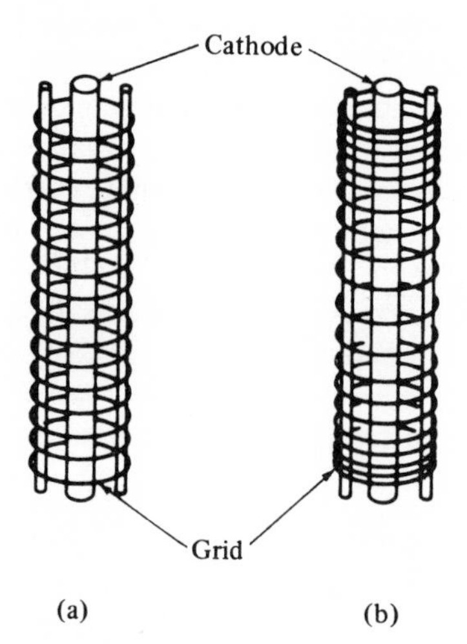

Figure 13–13 Grid structure for (a) ordinary tube and (b) variable-mu tube.

13-9 Gas-Filled Tubes

As an electron travels from cathode to anode it acquires considerable velocity and, therefore, a high kinetic energy. As a matter of fact this kinetic energy is somewhat of a problem in vacuum tubes because at impact the electrons strike so hard that some electrons are freed from the anode material.

It was discovered that if a controlled amount of a particular gas is placed within the glass envelope, the high-velocity electron will strike the gas molecules and in turn produce a similar, but more useful, phenomenon. The action is as follows. In a hot-cathode gas tube an electron is emitted by ordinary thermionic emission. Since the plate is positive, the electron heads toward the plate and increases its speed as it travels. If gas atoms are present in the proper amount, the electron strikes the gas atom. If the accelerated electron has sufficient energy, it will free an electron from the gas atom. This process of producing an electron and causing the original gas atom to be positive (short one electron) is called ionization. Now instead of one electron, there are two heading toward the plate. If the space between cathode and anode is large enough, the two electrons will acquire enough energy to produce two more electrons when collision occurs. If we multiply this one original electron by a large number, we can expect that the ionization process will be fast, and it will produce many electrons for external control. This ionization process can be recognized in tubes because it is accompanied by formation of light energy. Each type of gas gives its own characteristic color glow. Some of these gases are mercury vapor, neon, argon, and xenon.

A gas diode is used for rectification, and the advantage it has over vacuum diodes is that once ionization is achieved the voltage across a gas tube is considerably less than the voltage across a vacuum tube; see Figure 13-14. If both tubes were used in a rectifying circuit that required 200 mA, the voltage across the vacuum diode would be approximately 63 volts, but across the gas diode it would be only 16 volts. Since there is less voltage across the gas tube, more of the input voltage will appear across the series load resistance. This means that a higher dc output is realized; in other words, the gas tube is more efficient.

One word of caution is necessary as far as gas tubes are concerned. A series resistance must always be present whenever the tube is to conduct because once conduction starts, the current is limited by the external resistance. If there is no resistance or too low a resistance in series, the tube will burn out.

Another type of gas diode is the glow discharge tube, in which there is no

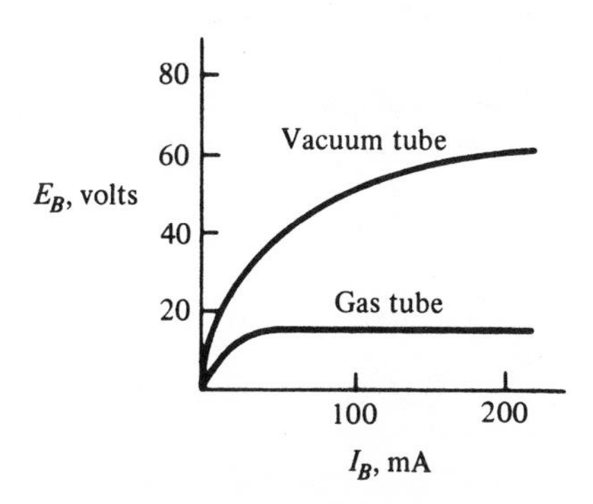

Figure 13–14 Comparison of voltage drops across vacuum diode and gas diode.

hot cathode. To start ionization, the plate-to-cathode voltage must be raised sufficiently so that it "tears" away an electron from the gas atom. Once this action starts, complete ionization takes place in a matter of microseconds. One major advantage of a cold-cathode tube is that the voltage across it remains relatively constant. See Figure 13-15. Note that a dot in the tube envelope is used to indicate a gas tube. If tube V_R is ionized, any fluctuations in V_{dc} will not appear across V_R. One must recall that a source voltage tends to distribute itself across series resistances according to the magnitude of these resistances. In addition, if the source voltage increases, this increase is also distributed. However, this is not the case in the type of circuit shown in Figure 13-15. A slight increase in the source, V_{DC}, will appear across R_S but not across V_R. The net result is that this type of tube is used in voltage regulation; that is, if variations in dc voltage are to be minimized, placing a cold-cathode tube in the circuit will regulate the output. This type of circuit is used in electronic equipment where it is necessary to keep a constant voltage. A typical example is in aircraft electronics. Here, a change in engine speed may reflect as a change in generator voltage output.

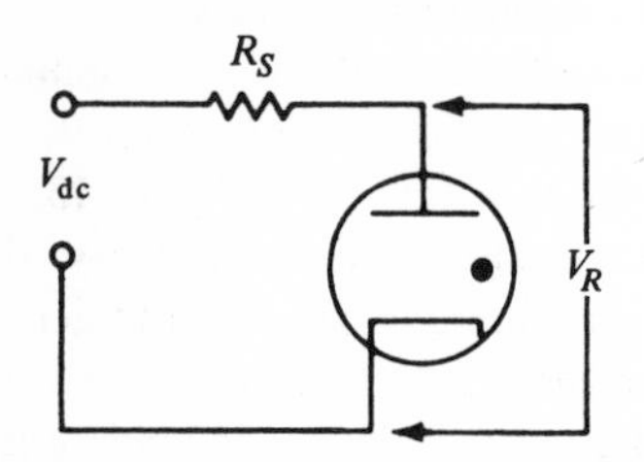

Figure 13–15 Cold-cathode gas diode used in a voltage-regulator circuit.

Manufacturers have produced different regulating voltages by using different gases and pressures. Standard voltage regulating voltages are 75, 105, and 150 volts. Combinations can be achieved by stacking these tubes in series. Caution must be observed using this approach because the current ratings of the individual tubes may be different.

It was inevitable that solid-state devices would replace gas regulators. Not only do they have the typical solid-state advantages, but Zeners have considerably more standard voltages; this, in turn, permits an even higher number of available voltages by series connection. Another disadvantage of gas-type tubes is that, since they are discharge devices, they generate radio-frequency energy. This energy radiates and may cause interference with other electronic circuits.

13-10 The Thyratron

The simplest way to describe a thyratron tube is to say it is a gas triode that behaves just like a silicon controlled rectifier (SCR). You recall that the SCR conducted current between anode and cathode provided the gate permitted conduction. If the gate current increased slightly, the anode–cathode conduction occurred at a lower voltage. The thyratron operates the same way, except that it is voltage-controlled; that is, the thyratron anode–cathode voltage will start conduction sooner if the grid bias voltage becomes less negative. A typical characteristic is shown in Figure 13-16(a). If the grid is at -9 volts, the anode voltage must exceed 500 volts before ionization, and hence conduction occurs. On the other hand, if the grid is -4 volts, conduction will begin when the plate is at 100 volts or more. Similarly, the grid loses control once the tube fires, just as the gate lost control in the SCR. You may recall from Chapter 8 that, in the SCR, regaining control is achieved by using ac anode voltage because ac returns to zero twice each cycle. The regaining of control is important because the amount of dc output can be controlled. Figure 13-16(b) shows a simplified circuit, where R_L is the load, E_{CC} the variable grid voltage, and ac is used as the source voltage. Despite the fact that the thyratron is a triode, it is not used as an amplifier but instead as a rectifier; the amount of dc is controllable. Figure 13-14(c) shows the

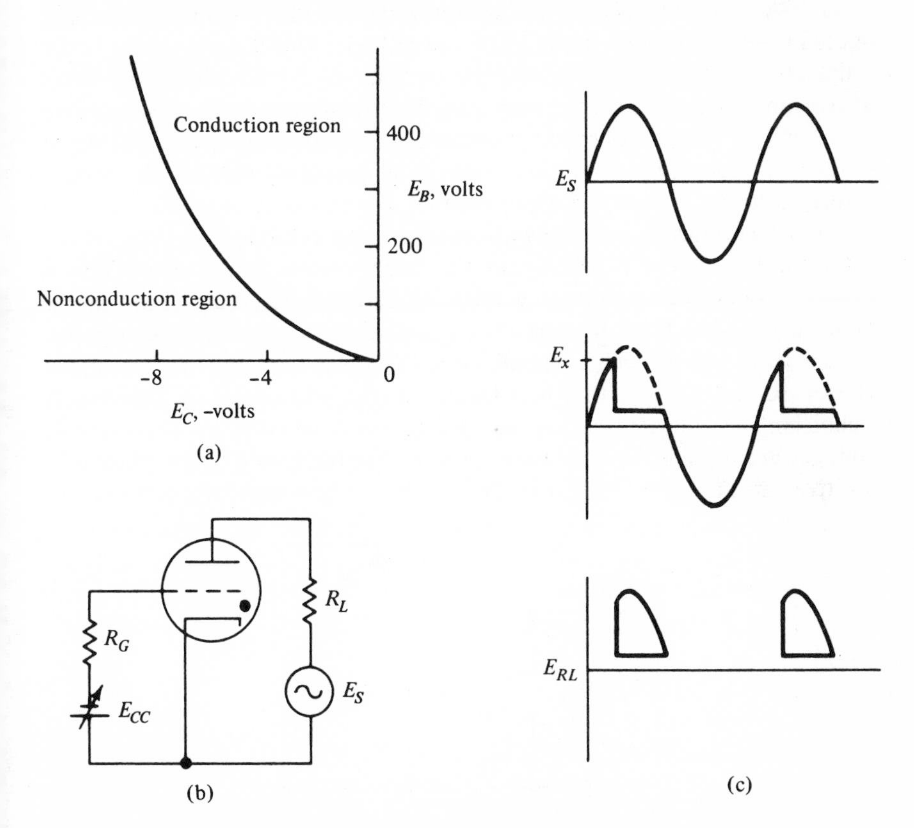

Figure 13–16 (a) Thyratron control characteristic. (b) Thyratron circuit with ac. (c) Waveforms of voltage input, voltage across tube, and voltage across R_L.

three important voltages that can be expected from the circuit of Figure 13-14(b). The source voltage E_S is applied to the series circuit R_L and the thyratron. For a given value of E_{CC} the tube will not conduct until a certain value, E_x, is reached. At that moment ionization takes place; the voltage across the tube drops and current flows. The voltage across the load is the remaining portion of the sine wave. When the source voltage drops to zero, current flow ceases, no voltage appears across R_L, and the grid regains control. During the negative half of the cycle, the anode is negative. Of course, no conduction occurs and the ions recombine to form atoms in a process called deionization. It is during this period that E_{CC} can be readjusted to produce less or more dc output across R_L. An increase in bias decreases the output; a decrease in bias increases the output.

13-11 The Phototube

A phototube is a diode (without a heater) that conducts when light of the proper wavelength impinges on the cathode. This is quite similar to the solid-state photodiode. Basically, the cathode is a half-cylinder of metal whose inner surface is coated with a light-sensitive material, such as cesium oxide, which releases electrons when enough light energy strikes it. As in solid-state photodevices, sufficient light of the proper wavelength is necessary before electrons are emitted. Once emitted, the electrons are attracted to an anode, which must be positive with respect to the cathode.

Phototube current is usually in the microampere range. Since this is usually insufficient to operate a relay or perform other useful work, the phototube circuit must be part of a circuit that produces larger currents. Figure 13-17 shows such a simple circuit. Notice both the symbol of the phototube and the fact that it is located in the grid circuit of a triode. A technician's circuit analysis can be used to check whether the necessary dc voltages are present for both tubes. E_{CC} makes the anode of the phototube positive and the cathode negative. Next, this voltage apparently serves as the

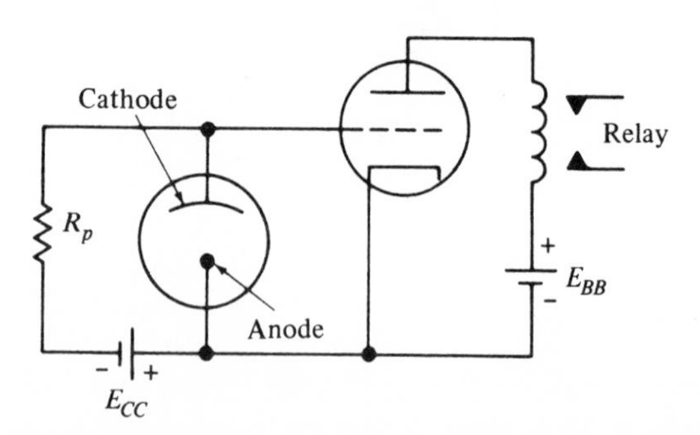

Figure 13–17 Light-sensitive circuit using phototube at input to a triode.

grid-to-cathode bias for the triode; if no light is present, the tube is not conducting and the grid indeed is negative with respect to the cathode and equal to E_{CC}. (The phototube is an open circuit and all the voltage appears across it.) Finally, the triode plate is made positive with respect to the cathode and will attract electrons, provided that the grid permits the passage.

Apparently, the dc voltages are appropriate and will make the tubes operable. Let us now investigate the effect of light. With the absence of light, all of E_{CC} appears between the grid and cathode. This makes the grid sufficiently negative to prohibit any electrons from reaching the plate. Thus, no current flows externally and the relay is not energized. If sufficient light at the proper wavelength strikes the phototube, electrons are emitted and current flows from E_{CC} through the phototube and R_p, and back to the battery. We are describing a simple series circuit where each resistance takes its share of the source voltage—in this case, E_{CC}. Therefore, the voltage across the phototube is somewhat less than E_{CC} and in a practical circuit the components are selected to produce a low voltage drop across the phototube. This lower voltage, which is also the bias for the triode, now permits electrons to reach the plate. These in turn yield plate current and thus energize the relay. If the light source is removed, the phototube does not conduct; thus, the triode bias becomes large (grid more negative), the plate current ceases, and the relay is deenergized.

It cannot be emphasized enough, as demonstrated in this simple circuit, that circuit analysis requires assurance that proper dc voltages are present. These voltages with proper polarity must permit conduction or, in other words, must have a complete circuit. In the phototube circuit, this includes the phototube and the triode. Analysis is completed by observing how each active electronic component behaves in the circuit with its own characteristic. Indeed, the chief intent of this text is to familiarize the student with the characteristics.

13-12 The Cathode-Ray Tube

One of the most useful pieces of equipment used in electronic work is the oscilloscope. Its primary purpose is to provide a means to *see* a given voltage as well as to indicate the magnitude. Obviously, the ordinary voltmeter measures magnitude only.

There are definite advantages in seeing a particular voltage. A common use, for instance, is in troubleshooting television sets. Certain voltage patterns appear as the signal travels through successive stages. These patterns are observed and followed. If the pattern appears different from that specified by the manufacturer, trouble apparently exists in the observed stage. The TV picture tube itself is another form of an oscilloscope except that the observed pattern is a series of pictures instead of voltage patterns.

The heart of these picture devices is the cathode-ray tube, shown in Figure 13-18. The essential parts of this type of tube are an electron gun, deflecting plates, and a fluorescent screen. The electron gun is similar to an ordinary vacuum tube in that it sends electrons from a cathode past a grid toward anodes. In this case, however, the cylindrical anodes do not stop these electrons. The holes in the anodes permit these accelerated electrons to proceed to a fluorescent screen at the face of this odd-shaped vacuum tube, which glows when electrons strike it. Just before the electron stream reaches the glowing material it passes between deflecting plates. One set can deflect the stream horizontally; the other set vertically. This task is relatively simple because electrons are negative and positive plates will exert an attracting influence as the electrons pass through the plate region. The net result is that the stream of electrons will strike the fluorescent material at a spot other than the center. Obviously, the amount of deflection depends on the amount of voltage on the plates. As shown in Figure 13-18 a positive upper plate with respect to the lower vertical plate causes the beam to produce a path shown as *A*. If the lower plate were made positive the beam would hit the screen near the bottom. Therefore, the spot can be positioned any place on the screen by a combination of vertical and horizontal voltages.

The oscilloscope makes use of this voltage-position phenomenon. On the horizontal deflecting plates a periodic voltage is placed so that the beam moves back and forth. However, the voltage is such that the stream is moved relatively slowly from left to right (looking at the screen) and very fast from right to left. The reason for this will be apparent shortly. The voltage to be observed is placed on the vertical plates and thus the stream will move up and down according to the magnitude of this voltage. The stream now moves up and down as well as across. Further adjustments on the oscilloscope permits the vertical-deflecting voltage to appear as a steady pattern by synchronizing the horizontal and vertical voltages. This means simply that every time the spot starts to move across, the vertical voltage has the same voltage it had the last time it appeared here. When the stream is told to

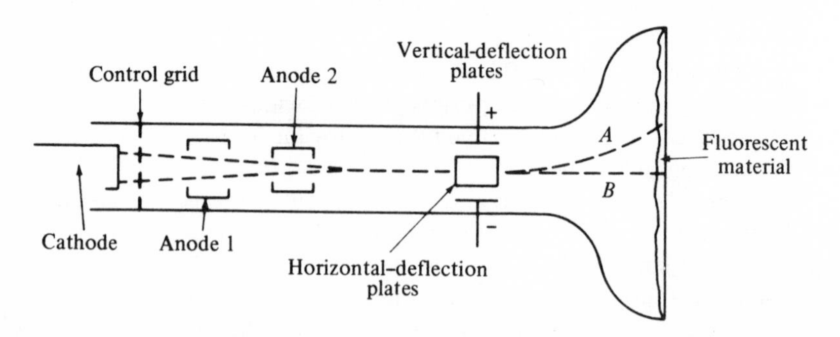

Figure 13–18 Essential parts of a cathode-ray tube.

return quickly to the left no picture or vertical deflection appears because the screen is automatically blanked out. It is as if one traced a slow sine wave continuously with chalk on a blackboard and when he reached the end of the board, returned to the same starting point very quickly. The result is that the same sine wave would be retraced every time. Essentially, television pictures are transmitted similarly, with the aid of synchronization.

Since the oscilloscope is probably the easiest exposure to a cathode-ray tube, some of the elementary controls are mentioned here so that some correlation can be established between this paragraph and the practical tangible oscilloscope. A first glance at a front panel tends to overwhelm an individual. However, no matter how complex oscilloscopes may be, they all have basic controls. For example, the intensity control varies the brightness of the spot or pattern that the screen makes. It controls the bias on the control grid. Focus control sharpens the screen image by adjusting one of the anode voltages. Position controls permit adjusting the spot or screen image vertically or horizontally. This is done by placing dc on the deflecting plates. The remaining controls are part of the synchronizing features of the oscilloscope. As an example, one control selects what will initiate the horizontal movement or trace. This knob has markings of internal, external, or line. Another control will determine how often the trace moves across the screen. This determines how many cycles (if periodic) of unknown voltage will appear across the face of the screen. If movement across the screen is relatively slow, many cycles of the unknown voltage will appear. In most applications an internal method of triggering is used and one or two cycles of signal are usually sufficient. In the vertical circuit two terminals are supplied to introduce the signal to be observed. An attenuation control nearby reduces relatively large signals before feeding them to the amplifier whose job it is to beef up small signals when they are to be observed.

Figure 13-19 is a simplification of the ac portion of the oscilloscope that was just described. Despite the simplification, it is believed sufficient to provide an elementary understanding and a basis for further, more detailed studies of the subject.

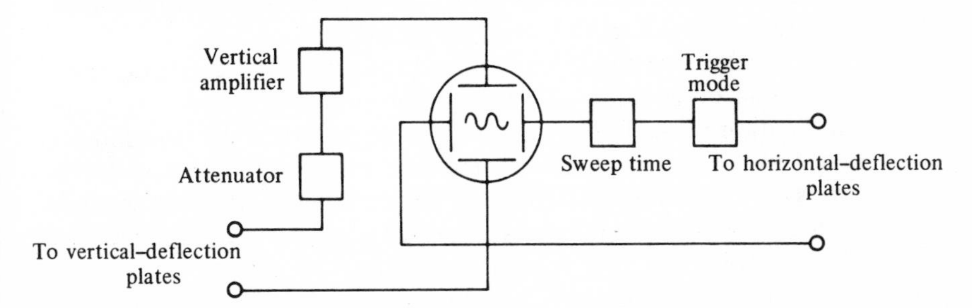

Figure 13–19 Simplified block diagram of the internal parts of an oscilloscope.

13-13 Summary

This chapter showed that vacuum tubes are similar to semiconductor diodes and transistors. The triode behaves like the transistor because electron flow, and then control, is established. The control effort is small in comparison to the output. This led to amplification of small signals.

The source of electrons in a vacuum tube is primarily achieved through heat. This type of emission is called thermionic. A positive anode attracts the electrons and placing a spiral-wound grid wire between the anode and cathode gives rise to control of the electron flow. Large negative grid bias reduces the electron flow to the anode. Varying signals on the grid, therefore, permit the electron stream to be modulated similarly.

We saw in this chapter how interelectrode capacitance hinders amplification of high frequencies. This problem led to the development of the pentode, a five-active-element tube, which has low interelectrode capacitance and characteristics similar to transistors.

In another type of vacuum tube investigated, the emission depends on light. This type is called photoemissive. The output current is relatively low and in general the phototube usually feeds another vacuum tube to strengthen the signal. And, as in solid-state photodevices, these photoemissive cathodes require the proper wavelength to energize them.

Gas-filled tubes were also studied. Ionization of gas atoms in the tube contributes many more electrons for attraction to the anode than in the vacuum tube; in addition, a peculiar phenomenon of constant tube voltage results. This constant voltage permits higher output voltages across the load; it is also useful in voltage-regulator applications. These tubes are called cold-cathode or glow-discharge gas tubes.

A three-element gas tube, called a thyratron, behaves like an SCR. However, the thyratron is a voltage-controlled device. Despite the three elements in the thyratron it is used as a rectifier and not as an amplifier. A convenient feature of this three-element device is that the amount of dc output can be controlled by a relatively small voltage on the grid.

The final vacuum tube introduced in this chapter was the cathode-ray tube. It permits visual observation of a voltage. In more sophisticated types of electronic hardware, visual observations are more important than magnitudes as obtained by voltmeters. Basically, the tube emits and focuses a stream of electrons onto a fluorescent screen. The stream is bent as it passes two sets of deflection plates—horizontal and vertical. To observe a given voltage on a cathode-ray oscilloscope, the voltage to be observed goes to the vertical deflecting plates. Simultaneously, a linear voltage is fed to the horizontal plates. The two are so synchronized that the observed voltage continuously describes the same path on the screen. The speed is so fast and the synchronization is so good that the pattern appears to be stationary.

Questions and Problems

1. What kind of energy is used to start conduction in most vacuum tubes?
2. What does work function describe?
3. (True or false.) All electron-emitting materials are placed on filaments.
4. (True or false.) The electron-emitting element is called the cathode.
5. (True or false.) The anode attracts electrons when it is positive with respect to the cathode.
6. (True or false.) A 4-volt change in control-grid voltage is equal to a 4-volt change in plate voltage on the amount of electrons reaching the plate.
7. Give the formula for (a) amplification factor, (b) transconductance, (c) ac plate resistance.
8. Which parameter in Problem 7 gives a theoretical voltage gain?
9. If a tube grid change of 2 volts requires a 20-volt change in plate voltage to keep the plate current constant, what is the amplification factor?
10. Determine the transconductance if a plate current change of 2 mA is caused by a grid voltage change of 4 volts when the plate voltage is held constant.
11. The plate current increases from 8 to 12 mA when the grid bias changes from -4 to -3 volts, with a constant plate voltage. What is the transconductance?
12. In Figure 13-6 determine μ if the tube operates at a plate voltage of 300 and a plate current of 1 mA.
13. Repeat Problem 13 for an operating point of $E_B = 160$ and $I_B =$ 12 mA. Compare with answer in Problem 12.
14. Obtain r_p at an operating point of Problem 12. ($E_B = 300$ and $I_B = 1$ mA.)
15. Obtain r_p at an operating point of Problem 13. ($E_B = 160$ and $I_B = 12$ mA.) Compare with Problem 14.
16. A tube has an r_p of 10 kΩ and a g_m of 1000 μmho. Determine its μ.
17. A tube has an r_p of 5.0 kΩ and a μ of 25. Determine g_m.
18. Determine g_m from the values obtained for r_p and μ in Problems 12 and 14.
19. Using Figure 13-6 and operating at point *B*, determine the change in plate voltage required to offset a change from -8 to -10 volts in grid voltage and thereby keep the plate current constant.
20. Tests on a particular tube produced the following data:

E_C	I_B	E_B
-6 V	10 mA	250 V
-8	8	250
-8	10	290

Determine g_m, r_p, and μ.

21. What factor limits high-frequency response in vacuum tubes?
22. Which type tube, pentode or triode, generally is better for high-frequency response? Why?
23. Which type tube has a characteristic negative slope?
24. What type of vacuum tube has characteristics that resemble a transistor *I-E* response?
25. What are the active elements in a pentode?
26. Why is a beam power tube so named?
27. What physically makes a remote-cutoff tube have a remote cutoff?
28. Indicate by use of a curve the difference between a variable-mu and an ordinary-mu tube.
29. Describe the action of a gas tube. What is the breakdown process called?
30. What is the chief advantage of a gas diode over a vacuum diode?
31. What are cold-cathode gas tubes used for? What characteristic do they possess that make them useful?
32. What is a thyratron tube? What makes it useful?
33. (True or false.) An increase in grid bias of a thyratron will require an increase in anode voltage to make it conduct.
34. (True or false.) Phototubes are relatively high-resistance devices even when they conduct.
35. (True or false.) Phototubes require filament voltage.
36. Name the three general parts of a cathode-ray tube.
37. What is the purpose of the electron gun?
38. What happens when an electron stream strikes the fluorescent material in the tube?
39. What mechanism is required to deflect the stream?
40. What advantage does the cathode-ray tube have over an ordinary voltmeter?
41. Describe how a sine wave can be portrayed on an oscilloscope.

Glossary

Glossary

Chapter 1

Proton: An atomic particle having a positive charge. The charge is equal in magnitude to the charge of an electron. The mass of the proton is 1850 times that of an electron and is located in the nuclei of matter.
Electron: An atomic particle having a negative charge. Electrons orbit the nucleus of a substance and are attracted to protons by an electrostatic force.
Electrostatic force: A force experienced by a particle when placed in an electric field.
Semiconductor: A class of solid materials characterized by relatively high resistivities which can be varied by special treatments. Some common semiconductors are silicon, germanium, and selenium.
Crystal: A solid in which the atoms are arranged in an orderly fashion (lattice) caused by the valence electrons in adjacent atoms.
Valence electrons: Electrons in the outer orbit of an atom. These electrons determine the relative activity of the atom with other atoms.
Impurity: A substance which when diffused into a semiconductor alters the latter's resistivity. An impurity that "releases" electrons is called a donor and one that "releases" holes is called an acceptor atom.
n-type crystal: A crystal that has donor impurity, thus making it "rich" in electrons.
p-type crystal: A crystal that has acceptor impurity and makes the crystal "rich" in holes or positive in nature.
p-n junction: The union between *p* and *n* layers in a single crystal.
Energy levels: Energy states of electrons in an atom. Electrons in the outer orbits have higher energy levels than those in inner orbits. To raise the energy level a precise amount of energy must be given to an electron.
Forbidden energy gap: A region between valence-energy levels and conduction-energy levels where no energy levels exist. To be conductive, this region, or gap, must be small. This means that little energy is added to valence electrons to make them become part of a conductive circuit.

Chapter 2

Potential barrier: A region in a semiconductor that prohibits electric charges from passing through. This is sometimes called a potential hill.
Reverse bias: dc voltage applied to a junction that increases the potential barrier yielding a low-conductivity, high-resistance semiconductor condition.
Leakage current: The current that flows when reverse bias is applied to a *p-n* junction. This is also called back current and usually consists of a few micro-amperes.
Forward bias: dc voltage applied to a junction that decreases the potential barrier permitting a high-conductivity, low-resistance condition.
Semiconductor diode: A single *p-n* junction that allows electrons to pass in one direction only.

Chapter 3

Transistor: A three-element semiconductor consisting of three conducting regions (*n-p-n* or *p-n-p*) in a single crystal. The three regions produce two junctions; one is forward-biased and the other is reverse-biased for normal operation.
Emitter, base, and collector: The three elements of a transistor that correspond to *p-n-p* or *n-p-n* doped regions in a transistor. In normal operation the emitter-base junction is forward-biased and the base collector is reverse-biased. Most of the emitter carriers travel through the base and are swept up by the collector.
Common base: A transistor circuit configuration where the base is common to input and output. The emitter is the input and collector is the output.
Alpha: A ratio between output current to input current I_C/I_E in a common-base circuit.
Common emitter: A transistor circuit where the emitter is common to input and output. The base is the input and the collector is the output.
Beta: A ratio between output current to input current $I_C I_B$ in a common-emitter circuit.
Common collector: A transistor circuit where the collector is common to input and output. The base is the input and the emitter is the output.

Chapter 4

Hybrid parameters: Operational characteristics for transistors that specifies its performance for small signals. There are four parameters quoted by manufacturers. These are used in determining total circuit performance in which the transistor is used.

dc beta: The dc current gain in a common-emitter circuit. It is defined as I_C/I_B. It is also identified as h_{FE}.
ac beta: The small signal current gain in a common-emitter circuit. Defined as i_C/i_B. It is also identified as one of the hybrid parameters, h_{fe}.

Chapter 5

Voltage gain: A ratio of signal output voltage to input voltage in a complete stage, or stages, containing the necessary biasing resistors, load, and capacitors.
Phase inversion: A 180° out-of-phase relationship between output and input when a periodic wave is amplified in a common-emitter amplifier.
Input resistance: The effective resistance that is offered by a transistor and associated circuit to an incoming signal. Each amplifier configuration offers its characteristic range of input resistance.
Bias stability: dc operating conditions designed into the circuit to minimize effects of temperature, and thus prevent thermal runaway.
Current gain: A ratio of signal output current to input current in a transistor stage or coupled stages.

Chapter 6

Field-effect transistor (FET): A transistor in which current flow affects an electric field that is transverse to the current flow.
Source drain and gate: The three elements in a field-effect transistor. The source-drain path is the main current flow. The gate-source junction controls the above current flow.
Channel: The region between gates where the main current passes and is affected by the gate potential.
Pinch-off voltage: The smallest value of gate voltage that "pinches off" the main source-drain current.
Transconductance: Conductance (I/E) that relates the output current to the input voltage. In an FET it is the ratio of a change in drain current to a change in gate voltage.
I_{DSS}: Drain current when the gate voltage is zero at a specified drain-source voltage.
Common drain: An FET configuration where the drain is common to input and output. This is also called the source follower.
Insulated gate (FET): An FET with a thin insulating layer between the gate and the FET channel. The characteristics are different than the regular (junction) FET.

Chapter 7

Unijunction transistor (UJT): A three-terminal semiconductor device whose characteristic differs considerably from a conventional transistor. Its chief characteristic is a negative-resistance response.
Negative resistance: A characteristic of some active devices where the current through decreases, as the voltage across increases. This portion of a response curve yields a negative slope.
Standoff ratio: A characteristic of a UJT that indicates the proportion of bias voltage expected at the emitter from the base-to-base voltage.

Chapter 8

Silicon controlled rectifier (SCR): A three-terminal device containing four doped regions (PNPN). The gate controls the firing point, and hence the amount of current that flows between the cathode and anode. Once conducting, the gate loses control and the anode-cathode voltage must return to zero to regain control.
Peak reverse voltage: The maximum continuous voltage that may be applied to a rectifying device, in reverse direction, without damaging it.
Holding current: The minimum current that must flow through the SCR, in the forward direction, to keep it in the conducting state.

Chapter 9

Diac: A bidirectional switching diode. Once proper anode-cathode voltage is reached, the diode conducts.
Triac: A bidirectional silicon-controlled rectifier. Proper gate current can trigger anode conduction with either gate or anode polarity.
Thyristor: Transistors that belong to a family of gate controlled or switch-like action for controlling relatively heavy loads.
Conduction angle: The number of electrical degrees of the applied ac voltage during which the thyristor is conducting.
Firing angle: The electrical angle at which the thyristor current is started.
Heat sinks: A metal that absorbs and reradiates heat from a hot object. High-power semiconductor devices require heat sinks. In some instances a metal chassis is used.
Thermal resistance: An indication of the ability of a device or substance to remove heat, expressed in degrees centigrade per watt. All units that are in thermal contact with a solid-state pellet must have sufficient heat dissipation ability. The most relied upon thermal resistance in power semi-conductors is the heat sink.

Chapter 10

Wavelength: The distance travelled by a wave in the time of one cycle. For light frequencies, it is determined by dividing the speed of light by frequency, and is quoted in units of angstroms.
Angstrom: A unit of wavelength of light where one angstrom equals 10^{-10} meters. The human eye is sensitive to light waves in the range of 3800 to 7600 angstroms.
Spectral response: The relative response of a photosensitive device to various wavelengths.
Photoconductive: Any device that changes its electrical conductivity with changes in illumination. Dark to light resistance ratio of 10,000/1 is obtainable.
Photovoltage: A light-sensitive device that produces a voltage when exposed to light. The most common type is the solar cell.
Solar cell: A device that converts sunlight into electric energy.
Photo emissive: Any substance that emits electrons when struck by light waves.
Photo diode: In solid state, it is a two-element photo-conductive device. It also exhibits photovoltaic characteristics.
Phototransistor: A transistor that uses light as an input signal. Because of its amplifying ability, the phototransistor is quite sensitive.

Chapter 11

Integrated circuit (IC): An electronic circuit contained in a single package wherein all the components are microminiturized. One advantage of IC's is that many similar circuits are made simultaneously on one wafer.
Semiconductor IC: An integrated circuit wherein all components such as transistors and resistors are fabricated on a common semiconductor substrate.
Thin-film IC: An integrated circuit wherein components are fabricated on a common insulating substrate such as glass.
Hybrid IC: A class of IC's where a combination of IC's are used, such as two semiconductor IC or a thin-film IC in combination with a semiconductor IC.
Mask: A screen that is placed over a substrate and permits etching, which is the depositing of materials through selected areas called windows.

Chapter 12

Zener diode: A semiconductor diode that has voltage-regulating ability. Regulation is achieved at the breakdown voltage which appears in the reverse-voltage region.

Breakdown or avalanche voltage: Reverse voltage at which the voltage across the zener diode remains constant.
p-i-n diode: A semiconductor device with an intrinsic region separating *p* and *n* regions. The resulting characteristic makes the *p-i-n* diode useful as a variable high frequency resistance.
Tunnel diode: A special diode that has a negative-resistance characteristic. This phenomenon occurs at low voltages wherein the electrons "tunnel" through the barrier. This type of characteristic makes the tunnel diode useful in switching and oscillator circuits.
Thermistor: A semiconductor resistor that has a large negative temperature coefficient of resistance. This temperature sensitivity makes it particularly adaptable to temperature measurement.
Varactor: A special semiconductor diode whose capacitance varies with voltage. This effect is caused by the removal of mobile carriers from the junction as reverse voltage is increased.
Light-emitting diodes (LED): A special semiconductor diode that emits light when a voltage is applied.
Hall-effect device: A device that has contacts on four edges of a special semiconductor wafer. If a voltage is applied across two terminals and a negative field appears perpendicular to the wafer, then a Hall voltage, proportional to the magnetic field and the impressed voltage, will appear across the other two terminals.

Chapter 13

Thermionic emission: The releasing of electrons from a surface by heating the surface.
Work function: The minimum energy required to cause an electron to be emitted from a metal surface. Usually expressed in electron volts or just volts.
Filament: The wire which current passes through in order to produce heat for electron emission.
Cathode: The electron-emitting electrode in a vacuum tube. The cathode may be directly or indirectly heated. When the former is true, the cathode is also the filament.
Anode: The element in vacuum tubes to which electrons are made to flow because the anode is made positive with respect to the cathode. It is frequently called the plate.
Vacuum diode: A two-element device that permits electrons to flow in one direction, from cathode to plate.
Child's law: An equation that states the relationship between current and voltage in a vacuum diode. Essentially the current is proportional to the three-halves power of the applied plate-cathode voltage.
Triode: A three-element vacuum tube that has a plate, cathode, and control

grid. The control grid is placed between the cathode and plate. When operated with a negative potential it tends to control the amount of electrons reaching the plate.

Amplification factor: A tube parameter that describes its theoretical voltage gain. It is defined as the ratio of a small change in plate voltage to a small change in grid voltage, with plate current constant.

Plate resistance: A tube parameter that describes the ac resistance the electrons encounter in flowing from cathode to plate. It is defined as a change in plate current while keeping the grid voltage constant.

Transconductance: A tube parameter that is defined as the ratio of a change in grid voltage, with the plate voltage constant.

Tetrode: A four-element vacuum tube where another grid, called a screen grid, is placed between the control grid and the plate.

Pentode: A five-element vacuum tube where another grid, called a suppressor grid, is placed between the screen grid and the plate.

Variable-mu tube: A vacuum tube with a grid design which causes an amplification that varies with the grid bias voltage. It is also called a remote cutoff tube.

Ionization: The breakup of a gas atom that results in a free electron and a positively charged ion.

Thyratron: A three-element gas tube containing a hot cathode, grid, and plate. Once conduction starts the grid loses control until the anode voltage returns to zero.

Cathode-ray tube: A special vacuum tube where electrons are shaped into a beam in the gun section and then accelerated to a fluorescent screen causing a spot of light to be emitted. This permits signals to be visible and arranged to show video signals as in television.

Index

Index

A

Acceptor, 8
Alpha, ac, 39
  dc, 22
Amplification, 18
Amplification factor, 205
Angstrom, 156
Atom, germanium, 3
  silicon, 3
Atomic number, 2

B

Back current, 15
Base, 18
Beta, ac, 36
  dc, 23
Biasing, simple, 80
  voltage divider-emitter resistance, 82
Bulk resistance, 30

C

Child's three-halves power law, 200
Collector, 18
Common base, 21, 32
Common collector, 24
Common emitter, 22
Common source, amplifier, 100
  gain, 102
Conduction angle, 147
Conduction band, 8
Control grid, 203
Coulombic force, 1
Covalent bond, 4
Current gain, 22

D

Depletion-enhancement mode, 109
Depletion region, 12
Derating curve, 150
Diac, 139
Diffusion, 6
Diode, light emitting, 192
  photo, 162
  p-i-n, 183
  semiconductor, 14
  tunnel, 184
  vacuum, 201
  zener, 180

Donor, 8
Doping, 7
Drift, 6

E

Electron, 1
Electronvolt, 8
Emitter, 18
Energy, kinetic, 2
  potential, 2
Energy level, 3
Epitaxial layer, 170

F

Field-effect transistor, 93
Filament, 199
Firing angle, 147
Flip-chip, 173
Forward bias, 13

G

g_m, 96, 206
g_{m0}, 99

H

Hall effect, 194
Heat sink, 149
Holding current, 129, 132
Holes, 5

I

I_{DSS}, 98
Input resistance, 71, 73
Insulated gate, FET, 109
Integrated circuit, compatible, 176
  hybrid, 175
  semiconductor, 169
  thin-film, 173

J

Junction, 12
Junction resistance, 31

L

λ, 155
Lattice, 5
Leakage current, 15, 39, 40
Light-emitting diodes, 192
Load line, 86

M

Majority carriers, 7
Micron, 156
Minority carriers, 7
Monolithic, 169
MOSFET, 109
μ, 205

N

n-channel, 94
Negative resistance, 115
Neutron, 2

O

Operating point, 86

P

p-channel, 94
Peak current, 117
  voltage, 117
Peak reverse voltage, 41
Pentavalent atom, 7
Phase shift, 148
Photoconductive, 158
Photodiode, 162
Phototransistor, 162

Photovoltaic, 160
Pinch-off, 94, 97
p-i-n diode, 183
Plate resistance, 206
Potential hill, 12
Proton, 2

R

r_{DS}, 97
r_p, 206
Reverse bias, 12

S

Silicon-controlled rectifier, 126
Solar cell, 161
Source follower, 105
Stage gain, 67
Standoff ratio, 117

T

Thermal resistance, 151
Thermionic emission, 198
Thermistor, 185
Thyristor triggering, 144
Transconductance, 96, 206
Triac, 139
Triode, 202
Trivalent atom, 8
Tunnel diode, 184

U

Unijunction transistor, 114
Unipolar, 95

V

V_{GS}, 95
Valance band, 8
Valley current, 117
  voltage, 117
Varactor, 188
Varistor, 187
Voltage gain, common-base, 61
  common-collector, 68
  common-emitter, 55
  FET, 98
  tube, 207

W

Wavelength, 155
Work function, 199

Z

Zener diode, 180